BRITANNICA

Mathematics in Context

Tracking Graphs

Britannica

ENCYCLOPÆDIA BRITANNICA EDUCATIONAL CORPORATION

Mathematics in Context is a comprehensive curriculum for the middle grades. It was developed in collaboration with the Wisconsin Center for Education Research, School of Education, University of Wisconsin–Madison and the Freudenthal Institute at the University of Utrecht, The Netherlands, with the support of National Science Foundation Grant No. 9054928.

National Science Foundation

Opinions expressed are those of the authors and not necessarily those of the Foundation

ISBN 0-7826-1512-0
1 2 3 4 5 WK 02 01 00 99 98

The *Mathematics in Context* Development Team

Mathematics in Context is a comprehensive curriculum for the middle grades. The National Science Foundation funded the National Center for Research in Mathematical Sciences Education at the University of Wisconsin–Madison to develop and field-test the materials from 1991 through 1996. The Freudenthal Institute at the University of Utrecht in The Netherlands, as a subcontractor, collaborated with the University of Wisconsin–Madison on the development of the curriculum.

The initial version of *Tracking Graphs* was developed by Jan Auke de Jong and Nanda Querelle. It was adapted for use in American schools by Margaret R. Meyer and Aaron N. Simon.

National Center for Research in Mathematical Sciences Education Staff

Thomas A. Romberg
Director

Joan Daniels Pedro
Assistant to the Director

Gail Burrill
Coordinator
Field Test Materials

Margaret R. Meyer
Coordinator
Pilot Test Materials

Mary Ann Fix
Editorial Coordinator

Sherian Foster
Editorial Coordinator

James A. Middleton
Pilot Test Coordinator

Margaret A. Pligge
First Edition Coordinator

Project Staff

Jonathan Brendefur
Laura J. Brinker
James Browne
Jack Burrill
Rose Byrd
Peter Christiansen
Barbara Clarke
Doug Clarke
Beth R. Cole
Fae Dremock
Jasmina Milinkovic
Kay Schultz
Mary C. Shafer
Julia A. Shew
Aaron N. Simon
Marvin Smith
Stephanie Z. Smith
Mary S. Spence
Kathleen A. Steele

Freudenthal Institute Staff

Jan de Lange
Director

Els Feijs
Coordinator

Martin van Reeuwijk
Coordinator

Project Staff

Mieke Abels
Nina Boswinkel
Frans van Galen
Koeno Gravemeijer
Marja van den Heuvel-Panhuizen
Jan Auke de Jong
Vincent Jonker
Ronald Keijzer
Martin Kindt
Jansie Niehaus
Nanda Querelle
Anton Roodhardt
Leen Streefland
Adri Treffers
Monica Wijers
Astrid de Wild

Acknowledgments

Several school districts used and evaluated one or more versions of the materials: Ames Community School District, Ames, Iowa; Parkway School District, Chesterfield, Missouri; Stoughton Area School District, Stoughton, Wisconsin; Madison Metropolitan School District, Madison, Wisconsin; Milwaukee Public Schools, Milwaukee, Wisconsin; and Dodgeville School District, Dodgeville, Wisconsin. Two sites were involved in staff developments as well as formative evaluation of materials: Culver City, California, and Memphis, Tennessee. Two sites were developed through partnership with Encyclopædia Britannica Educational Corporation: Miami, Florida, and Puerto Rico. University Partnerships were developed with mathematics educators who worked with preservice teachers to familiarize them with the curriculum and to obtain their advice on the curriculum materials. The materials were also used at several other schools throughout the United States.

We at Encyclopædia Britannica Educational Corporation extend our thanks to all who had a part in making this program a success. Some of the participants instrumental in the program's development are as follows:

Allapattah Middle School
Miami, Florida
Nemtalla (Nikolai) Barakat

Ames Middle School
Ames, Iowa
Kathleen Coe
Judd Freeman
Gary W. Schnieder
Ronald H. Stromen
Lyn Terrill

Bellerive Elementary
Creve Coeur, Missouri
Judy Hetterscheidt
Donna Lohman
Gary Alan Nunn
Jakke Tchang

Brookline Public Schools
Brookline, Massachusetts
Rhonda K. Weinstein
Deborah Winkler

Cass Middle School
Milwaukee, Wisconsin
Tami Molenda
Kyle F. Witty

Central Middle School
Waukesha, Wisconsin
Nancy Reese

Craigmont Middle School
Memphis, Tennessee
Sharon G. Ritz
Mardest K. VanHooks

Crestwood Elementary
Madison, Wisconsin
Diane Hein
John Kalson

Culver City Middle School
Culver City, California
Marilyn Culbertson
Joel Evans
Joy Ellen Kitzmiller
Patricia R. O'Connor
Myrna Ann Perks, Ph.D.
David H. Sanchez
John Tobias
Kelley Wilcox

Cutler Ridge Middle School
Miami, Florida
Lorraine A. Valladares

Dodgeville Middle School
Dodgeville, Wisconsin
Jacqueline A. Kamps
Carol Wolf

Edwards Elementary
Ames, Iowa
Diana Schmidt

Fox Prairie Elementary
Stoughton, Wisconsin
Tony Hjelle

Grahamwood Elementary
Memphis, Tennessee
M. Lynn McGoff
Alberta Sullivan

Henry M. Flagler Elementary
Miami, Florida
Frances R. Harmon

Horning Middle School
Waukesha, Wisconsin
Connie J. Marose
Thomas F. Clark

Huegel Elementary
Madison, Wisconsin
Nancy Brill
Teri Hedges
Carol Murphy

Hutchison Middle School
Memphis, Tennessee
Maria M. Burke
Vicki Fisher
Nancy D. Robinson

Idlewild Elementary
Memphis, Tennessee
Linda Eller

Jefferson Elementary
Santa Ana, California
Lydia Romero-Cruz

Jefferson Middle School
Madison, Wisconsin
Jane A. Beebe
Catherine Buege
Linda Grimmer
John Grueneberg
Nancy Howard
Annette Porter
Stephen H. Sprague
Dan Takkunen
Michael J. Vena

Jesus Sanabria Cruz School
Yabucoa, Puerto Rico
Andreíta Santiago Serrano

John Muir Elementary School
Madison, Wisconsin
Julie D'Onofrio
Jane M. Allen-Jauch
Kent Wells

Kegonsa Elementary
Stoughton, Wisconsin
Mary Buchholz
Louisa Havlik
Joan Olsen
Dominic Weisse

Linwood Howe Elementary
Culver City, California
Sandra Checel
Ellen Thireos

Mitchell Elementary
Ames, Iowa
Henry Gray
Matt Ludwig

New School of Northern Virginia
Fairfax, Virginia
Denise Jones

Northwood Elementary
Ames, Iowa
Eleanor M. Thomas

Orchard Ridge Elementary
Madison, Wisconsin
Mary Paquette
Carrie Valentine

Parkway West Middle School
Chesterfield, Missouri
Elissa Aiken
Ann Brenner
Gail R. Smith

Ridgeway Elementary
Ridgeway, Wisconsin
Lois Powell
Florence M. Wasley

Roosevelt Elementary
Ames, Iowa
Linda A. Carver

Roosevelt Middle
Milwaukee, Wisconsin
Sandra Simmons

Ross Elementary
Creve Coeur, Missouri
Annette Isselhard
Sheldon B. Korklan
Victoria Linn
Kathy Stamer

St. Joseph's School
Dodgeville, Wisconsin
Rita Van Dyck
Sharon Wimer

St. Maarten Academy
St. Peters, St. Maarten, NA
Shareed Hussain

Sarah Scott Middle School
Milwaukee, Wisconsin
Kevin Haddon

Sawyer Elementary
Ames, Iowa
Karen Bush Hoiberg

Sennett Middle School
Madison, Wisconsin
Brenda Abitz
Lois Bell
Shawn M. Jacobs

Sholes Middle School
Milwaukee, Wisconsin
Chris Gardner
Ken Haddon

Stephens Elementary
Madison, Wisconsin
Katherine Hogan
Shirley M. Steinbach
Kathleen H. Vegter

Stoughton Middle School
Stoughton, Wisconsin
Sally Bertelson
Polly Goepfert
Jacqueline M. Harris
Penny Vodak

Toki Middle School
Madison, Wisconsin
Gail J. Anderson
Vicky Grice
Mary M. Ihlenfeldt
Steve Jernegan
Jim Leidel
Theresa Loehr
Maryann Stephenson
Barbara Takkunen
Carol Welsch

Trowbridge Elementary
Milwaukee, Wisconsin
Jacqueline A. Nowak

W. R. Thomas Middle School
Miami, Florida
Michael Paloger

Wooddale Elementary Middle School
Memphis, Tennessee
Velma Quinn Hodges
Jacqueline Marie Hunt

Yahara Elementary
Stoughton, Wisconsin
Mary Bennett
Kevin Wright

Site Coordinators

Mary L. Delagardelle—Ames Community Schools, Ames, Iowa

Dr. Hector Hirigoyen—Miami, Florida

Audrey Jackson—Parkway School District, Chesterfield, Missouri

Jorge M. López—Puerto Rico

Susan Militello—Memphis, Tennessee

Carol Pudlin—Culver City, California

Reviewers and Consultants

Michael N. Bleicher
Professor of Mathematics
University of Wisconsin–Madison
Madison, WI

Diane J. Briars
Mathematics Specialist
Pittsburgh Public Schools
Pittsburgh, PA

Donald Chambers
Director of Dissemination
University of Wisconsin–Madison
Madison, WI

Don W. Collins
Assistant Professor of Mathematics Education
Western Kentucky University
Bowling Green, KY

Joan Elder
Mathematics Consultant
Los Angeles Unified School District
Los Angeles, CA

Elizabeth Fennema
Professor of Curriculum and Instruction
University of Wisconsin–Madison
Madison, WI

Nancy N. Gates
University of Memphis
Memphis, TN

Jane Donnelly Gawronski
Superintendent
Escondido Union High School
Escondido, CA

M. Elizabeth Graue
Assistant Professor of Curriculum and Instruction
University of Wisconsin–Madison
Madison, WI

Jodean E. Grunow
Consultant
Wisconsin Department of Public Instruction
Madison, WI

John G. Harvey
Professor of Mathematics and Curriculum & Instruction
University of Wisconsin–Madison
Madison, WI

Simon Hellerstein
Professor of Mathematics
University of Wisconsin–Madison
Madison, WI

Elaine J. Hutchinson
Senior Lecturer
University of Wisconsin–Stevens Point
Stevens Point, WI

Richard A. Johnson
Professor of Statistics
University of Wisconsin–Madison
Madison, WI

James J. Kaput
Professor of Mathematics
University of Massachusetts–Dartmouth
Dartmouth, MA

Richard Lehrer
Professor of Educational Psychology
University of Wisconsin–Madison
Madison, WI

Richard Lesh
Professor of Mathematics
University of Massachusetts–Dartmouth
Dartmouth, MA

Mary M. Lindquist
Callaway Professor of Mathematics Education
Columbus College
Columbus, GA

Baudilio (Bob) Mora
Coordinator of Mathematics & Instructional Technology
Carrollton-Farmers Branch Independent School District
Carrollton, TX

Paul Trafton
Professor of Mathematics
University of Northern Iowa
Cedar Falls, IA

Norman L. Webb
Research Scientist
University of Wisconsin–Madison
Madison, WI

Paul H. Williams
Professor of Plant Pathology
University of Wisconsin–Madison
Madison, WI

Linda Dager Wilson
Assistant Professor
University of Delaware
Newark, DE

Robert L. Wilson
Professor of Mathematics
University of Wisconsin–Madison
Madison, WI

TABLE OF CONTENTS

Britannica Mathematics in Context

Dear Teacher,

Welcome! *Mathematics in Context* is designed to reflect the National Council of Teachers of Mathematics Standards for School Mathematics and to ground mathematical content in a variety of real-world contexts. Rather than relying on you to explain and demonstrate generalized definitions, rules, or algorithms, students investigate questions directly related to a particular context and construct mathematical understanding and meaning from that context.

The curriculum encompasses 10 units per grade level. *Tracking Graphs* is designed to be one of four units in the algebra strand for grade 6/7, but it also lends itself to independent use. Students will produce line graphs from stories or data, and learn to look globally at line graphs.

In addition to the Teacher Guide and Student Books, *Mathematics in Context* offers the following components that will inform and support your teaching:

- *Teacher Resource and Implementation Guide,* which provides an overview of the complete system, including program implementation, philosophy, and rationale
- *Number Tools,* Volumes 1 and 2, which are a series of blackline masters that serve as review sheets or practice pages involving number issues/basic skills
- *News in Numbers,* which is a set of additional activities that can be inserted between or within other units; it includes a number of measurement problems that require estimation.

Thank you for choosing *Mathematics in Context.* We wish you success and inspiration!

Sincerely,

The Mathematics in Context Development Team

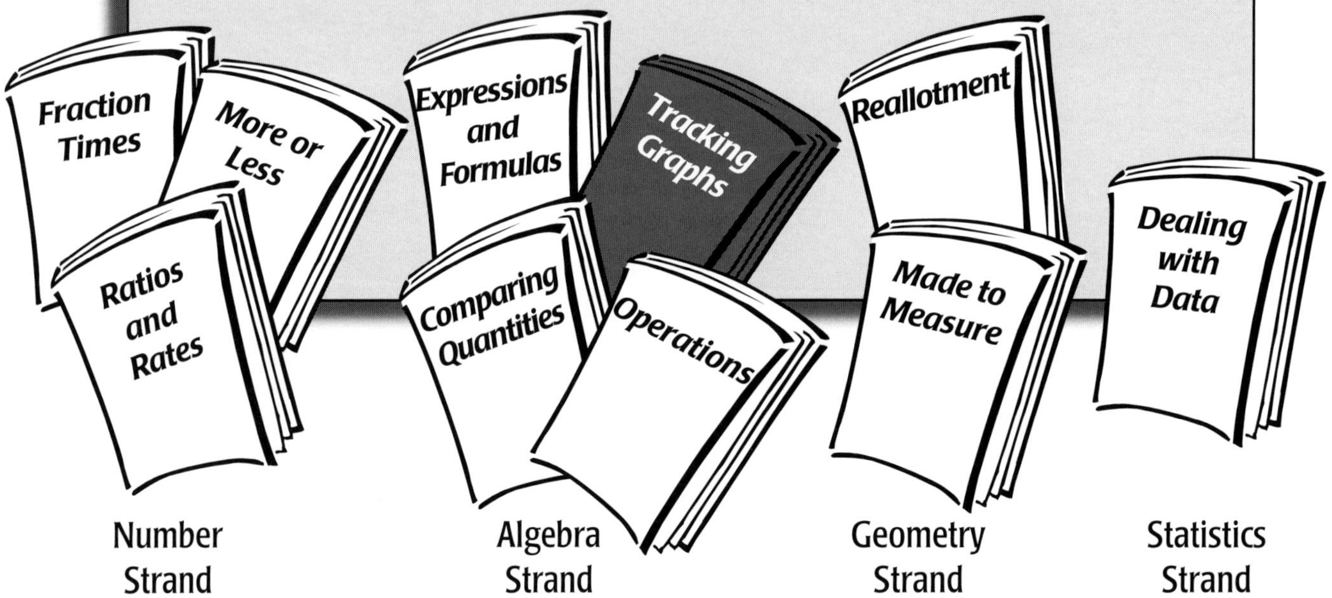

Overview

+15 min +65 min +15 min +15 m
+1 hour
+1 hour
+1 hour
BUS SCHEDULE
OUTDOOR
96.8
77.5
INDOOR
BRITANNICA
Mathematics in Context
Gino's

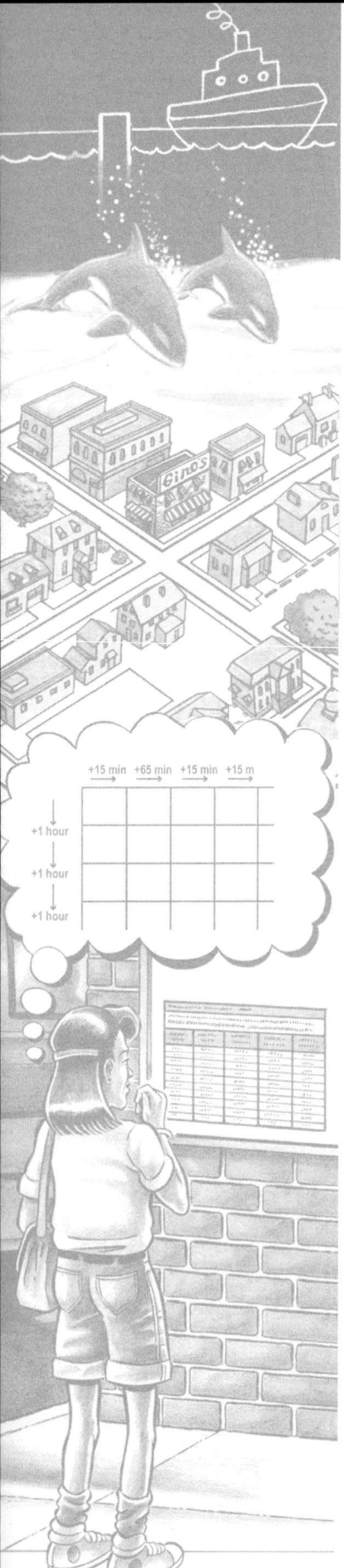

How to Use This Book

This unit is one of 40 for the middle grades. Each unit can be used independently; however, the 40 units are designed to make up a complete, connected curriculum (10 units per grade level). There is a Student Book and a Teacher Guide for each unit.

Each Teacher Guide comprises elements that assist the teacher in the presentation of concepts and in understanding the general direction of the unit and the program as a whole. Becoming familiar with this structure will make using the units easier.

Each Teacher Guide consists of six basic parts:

- Overview
- Student Materials and Teaching Notes
- Assessment Activities and Solutions
- Glossary
- Blackline Masters
- Try This! Solutions

Overview

Before beginning this unit, read the Overview in order to understand the purpose of the unit and to develop strategies for facilitating instruction. The Overview provides helpful information about the unit's focus, pacing, goals, and assessment, as well as explanations about how the unit fits with the rest of the *Mathematics in Context* curriculum.

Student Materials and Teaching Notes

This Teacher Guide contains all of the student pages (except the Try This! activities), each of which faces a page of solutions, samples of students' work, and hints and comments about how to facilitate instruction. Note: Solutions for the Try This! activities can be found at the back of this Teacher Guide.

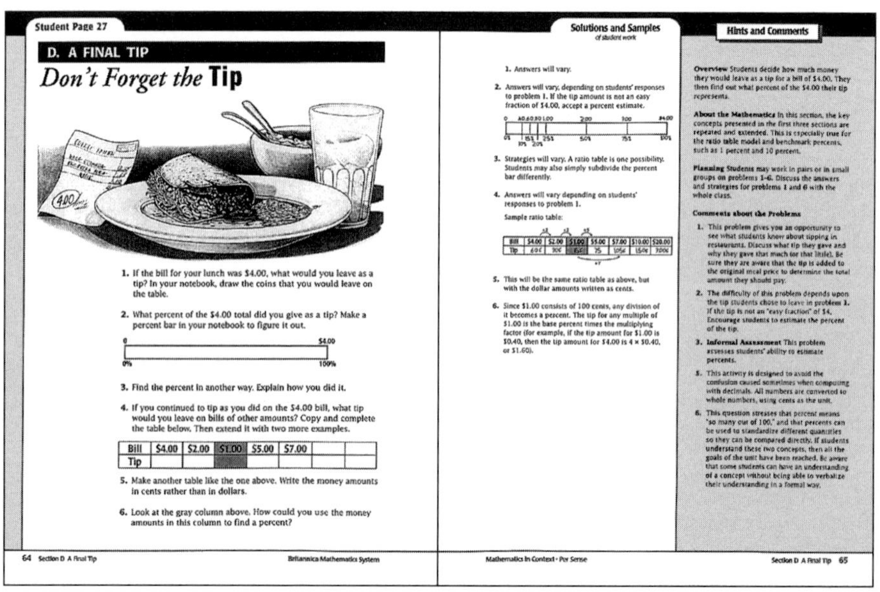
Student Page 27

D. A FINAL TIP

Don't Forget the **Tip**

1. If the bill for your lunch was $4.00, what would you leave as a tip? In your notebook, draw the coins that you would leave on the table.
2. What percent of the $4.00 total did you give as a tip? Make a percent bar in your notebook to figure it out.

0 — $4.00
0% — 100%

3. Find the percent in another way. Explain how you did it.
4. If you continued to tip as you did on the $4.00 bill, what tip would you leave on bills of other amounts? Copy and complete the table below. Then extend it with two more examples.

Bill	$4.00	$2.00	$1.00	$5.00	$7.00		
Tip							

5. Make another table like the one above. Write the money amounts in cents rather than in dollars.
6. Look at the gray column above. How could you use the money amounts in this column to find a percent?

64 Section D A Final Tip — Britannica Mathematics System

Solutions and Samples
of student work

1. Answers will vary.
2. Answers will vary, depending on students' responses to problem 1. If the tip amount is not an easy fraction of $4.00, accept a percent estimate.
3. Strategies will vary. A ratio table is one possibility. Students may also simply subdivide the percent bar differently.
4. Answers will vary depending on students' responses to problem 1.

Sample ratio table:

Bill	$4.00	$2.00	$1.00	$5.00	$7.00	$10.00	$20.00
Tip	60¢	30¢	15¢	75	105¢	150¢	300¢

5. This will be the same ratio table as above, but with the dollar amounts written as cents.
6. Since $1.00 consists of 100 cents, any division of it becomes a percent. The tip for any multiple of $1.00 is the base percent times the multiplying factor (for example, if the tip amount for $1.00 is $0.40, then the tip amount for $4.00 is 4 × $0.40, or $1.60).

Mathematics in Context • Per Sense

Hints and Comments

Overview Students decide how much money they would leave as a tip for a bill of $4.00. They then find out what percent of the $4.00 their tip represents.

About the Mathematics In this section, the key concepts presented in the first three sections are repeated and extended. This is especially true for the ratio table model and benchmark percents, such as 1 percent and 10 percent.

Planning Students may work in pairs or in small groups on problems 1–6. Discuss the answers and strategies for problems 1 and 6 with the whole class.

Comments about the Problems

1. This problem gives you an opportunity to see what students know about tipping in restaurants. Discuss what tip they gave and why they gave that much (or that little). Be sure they are aware that the tip is added to the original meal price to determine the total amount they should pay.
2. The difficulty of this problem depends upon the tip students chose to leave in problem 1. If the tip is not an "easy fraction" of $4, Encourage students to estimate the percent of the tip.
3. **Informal Assessment** This problem assesses students' ability to estimate percents.
5. This activity is designed to avoid the confusion caused sometimes when computing with decimals. All numbers are converted to whole numbers, using cents as the unit.
6. This question stresses that percent means "so many out of 100," and that percents can be used to standardize different quantities so they can be compared directly. If students understand these two concepts, then all the goals of the unit have been reached. Be aware that some students can have an understanding of a concept without being able to verbalize their understanding in a formal way.

Section D A Final Tip 65

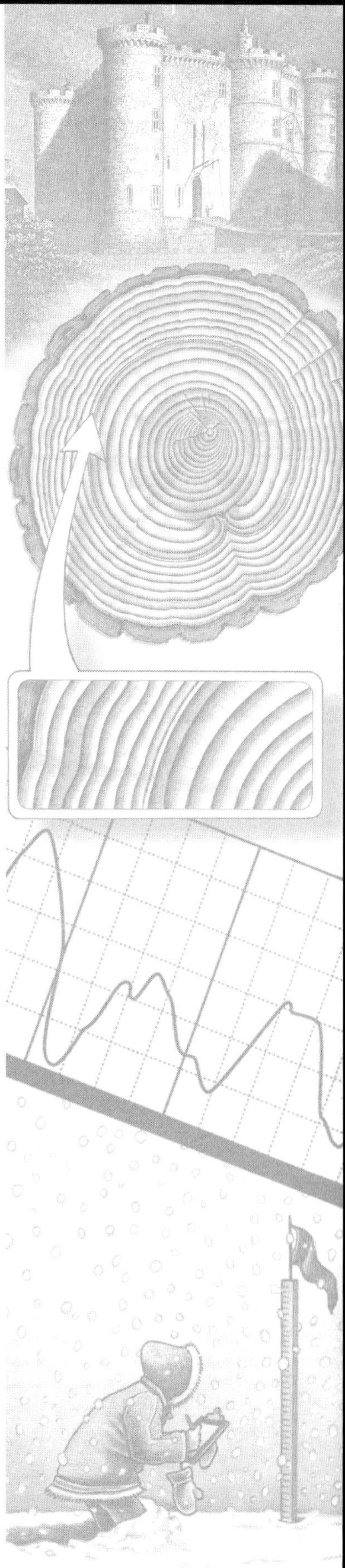

Each section within the unit begins with a two-page spread that describes the work students do, the goals of the section, new vocabulary, and materials needed, as well as providing information about the mathematics in the section and ideas for pacing, planning instruction, homework, and assessment.

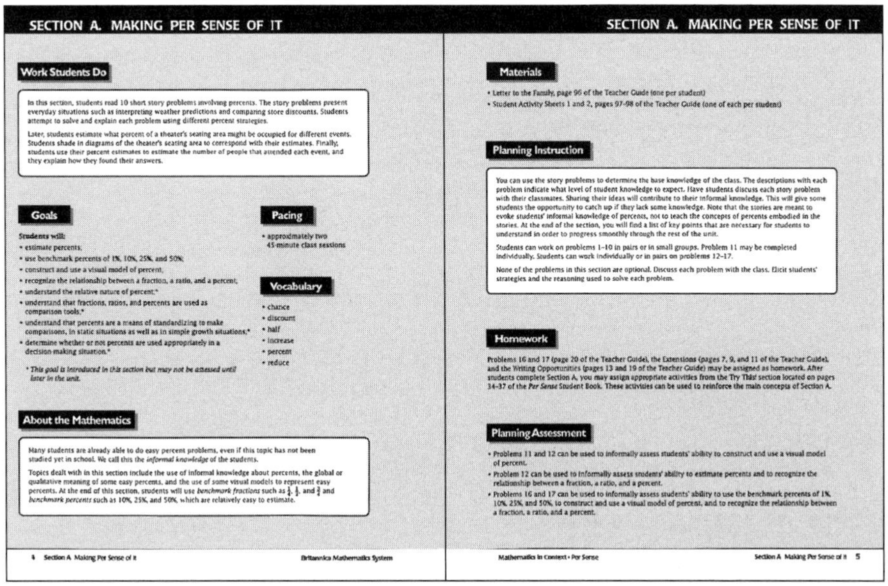

SECTION A. MAKING PER SENSE OF IT

Work Students Do

In this section, students read 10 short story problems involving percents. The story problems present everyday situations such as interpreting weather predictions and comparing store discounts. Students attempt to solve and explain each problem using different percent strategies.

Later, students estimate what percent of a theater's seating area might be occupied for different events. Students shade in diagrams of the theater's seating area to correspond with their estimates. Finally, students use their percent estimates to estimate the number of people that attended each event, and they explain how they found their answers.

Goals

Students will:

- estimate percents;
- use benchmark percents of 1%, 10%, 25%, and 50%;
- construct and use a visual model of percent;
- recognize the relationship between a fraction, a ratio, and a percent;
- understand the relative nature of percent;*
- understand that fractions, ratios, and percents are used as comparison tools;*
- understand that percents are a means of standardizing to make comparisons, in static situations as well as in simple growth situations;*
- determine whether or not percents are used appropriately in a decision making situation.*

** This goal is introduced in this section but may not be assessed until later in the unit.*

Pacing

- approximately two 45-minute class sessions

Vocabulary

- chance
- discount
- half
- increase
- percent
- reduce

About the Mathematics

Many students are already able to do easy percent problems, even if this topic has not been studied yet in school. We call this the *informal knowledge* of the students.

Topics dealt with in this section include the use of informal knowledge about percents, the global or qualitative meaning of some easy percents, and the use of some visual models to represent easy percents. At the end of this section, students will use *benchmark fractions* such as $\frac{1}{4}$, $\frac{1}{2}$, and $\frac{3}{4}$ and *benchmark percents* such as 10%, 25%, and 50%, which are relatively easy to estimate.

4 Section A Making Per Sense of It   Britannica Mathematics System

SECTION A. MAKING PER SENSE OF IT

Materials

- Letter to the Family, page 96 of the Teacher Guide (one per student)
- Student Activity Sheets 1 and 2, pages 97–98 of the Teacher Guide (one of each per student)

Planning Instruction

You can use the story problems to determine the base knowledge of the class. The descriptions with each problem indicate what level of student knowledge to expect. Have students discuss each story problem with their classmates. Sharing their ideas will contribute to their informal knowledge. This will give some students the opportunity to catch up if they lack some knowledge. Note that the stories are meant to evoke students' informal knowledge of percents, not to teach the concepts of percents embodied in the stories. At the end of the section, you will find a list of key points that are necessary for students to understand in order to progress smoothly through the rest of the unit.

Students can work on problems 1–10 in pairs or in small groups. Problem 11 may be completed individually. Students can work individually or in pairs on problems 12–17.

None of the problems in this section are optional. Discuss each problem with the class. Elicit students' strategies and the reasoning used to solve each problem.

Homework

Problems 16 and 17 (page 20 of the Teacher Guide), the Extensions (pages 7, 9, and 11 of the Teacher Guide), and the Writing Opportunities (pages 13 and 19 of the Teacher Guide) may be assigned as homework. After students complete Section A, you may assign appropriate activities from the Try This! section located on pages 34–37 of the *Per Sense* Student Book. These activities can be used to reinforce the main concepts of Section A.

Planning Assessment

- Problems 11 and 12 can be used to informally assess students' ability to construct and use a visual model of percent.
- Problem 12 can be used to informally assess students' ability to estimate percents and to recognize the relationship between a fraction, a ratio, and a percent.
- Problems 16 and 17 can be used to informally assess students' ability to use the benchmark percents of 1%, 10%, 25%, and 50%, to construct and use a visual model of percent, and to recognize the relationship between a fraction, a ratio, and a percent.

Mathematics in Context • Per Sense   Section A Making Per Sense of It 5

Assessment Activities and Solutions

Information about assessment can be found in several places in this Teacher Guide. General information about assessment is given in the Overview; informal assessment opportunities are identified on the teacher pages that face each student page; and the Assessment Activities section of this guide provides formal assessment opportunities.

Glossary

The Glossary defines all vocabulary words listed on the Section Opener pages. It includes the mathematical terms that may be new to students, as well as words associated with the contexts introduced in the unit. (Note: The Student Book does not have a glossary. This allows students to construct their own definitions, based on their personal experiences with the unit activities.)

Blackline Masters

At the back of this Teacher Guide are blackline masters for photocopying. The blackline masters include a letter to families (to be sent home with students before beginning the unit), several student activity sheets, and assessment masters.

Try This! Solutions

Also included in the back of this Teacher Guide are the solutions to several Try This! activities—one related to each section of the unit—that can be used to reinforce the unit's main concepts. The Try This! activities are located in the back of the Student Book.

Unit Focus

In this unit, students look at line graphs in two ways: locally and globally. They work locally, explaining individual points in a diagram of a race and in a graph of temperature changes. Students are gradually challenged to take a more global look at graphs, focusing on essential features like shape, decrease and increase, and so on, rather than on precise details.

The contexts in this unit are all represented as functions of time, so all graphs are continuous.

Mathematical Content

- representing data with a graph
- extracting data from graphs and using the data to draw conclusions
- looking globally at graphs
- changing the horizontal scale in graphs
- translating line graphs (in vertical direction)

Prior Knowledge

This unit assumes students can already add and subtract decimal numbers. It will also be helpful if students have experience with rounding and estimating numbers using scale lines (see the grade 5/6 unit *Grasping Sizes*), experience with graphs (see the grade 5/6 unit *Picturing Numbers*), experience with temperature in degrees Fahrenheit and Celsius, and an understanding of oceanic tides.

Planning and Preparation

Pacing: 13 days

Section	Work Students Do	Pacing*	Materials
A. Graphing a Race	■ analyze a graphical representation of a 100-meter Olympic race ■ create a line graph of the data from a race between two runners ■ learn how to make line graphs of distance over time ■ analyze other races, create graphs, and analyze given graphical representations	3 days	■ Letter to the Family (one per student) ■ Student Activity Sheets 1, 2, and 4 (one of each per student) ■ Student Activity Sheet 3 (two per student) ■ See page 5 of the Teacher Guide for a complete list of the materials and the quantities needed.
B. Filling in Temperatures	■ make a line graph of temperatures over time ■ look globally at line graphs by matching verbal descriptions to graphs	4 days	■ Student Activity Sheets 5–9 (one of each per student) ■ See page 27 of the Teacher Guide for a complete list of the materials and the quantities needed.
C. Graphs over a Longer Time Period	■ compress a horizontal axis ■ see how changing the scale affects the graph ■ examine how compressing the time scale causes a loss of detail	2 days	■ Student Activity Sheets 10–12 (one of each per student) ■ See page 49 of the Teacher Guide for a complete list of the materials and the quantities needed.
D. Graphs Depicting Speed	■ create line graphs of speed over time ■ see a line graph as the record of a quantity changing over time ■ analyze complicated speed graphs and create one detailing a personal trip	2 days	■ See page 67 of the Teacher Guide for a complete list of the materials and the quantities needed.
E. Graphs of Tides	■ investigate tides and use the information to make decisions on whether ships can enter a harbor or not ■ produce and examine a vertical translation or shift of a graph	2 days	■ Student Activity Sheets 13 and 14 (one of each per student) ■ See page 77 of the Teacher Guide for a complete list of the materials and the quantities needed.

* One day is approximately equivalent to one 45-minute class session.

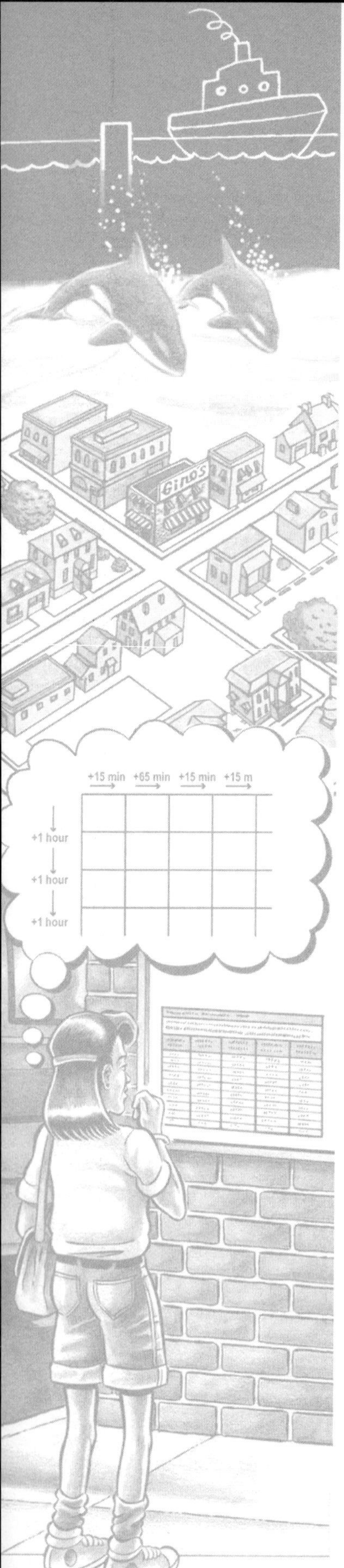

Preparation

In the *Teacher Resource and Implementation Guide* is an extensive description of the philosophy underlying both the content and the pedagogy of the *Mathematics in Context* curriculum. Suggestions for preparation are also given in the Hints and Comments columns of this Teacher Guide. You may want to consider the following:

- Work through the unit before teaching it. If possible, take on the role of the student and discuss your strategies with other teachers.
- Use the overhead projector for student demonstrations, particularly with overhead transparencies of the student activity sheets and any manipulatives used in the unit.
- Invite students to use drawings and examples to illustrate and clarify their answers.
- Allow students to work at different levels of sophistication. Some students may need concrete materials, while others can work at a more abstract level.
- Provide opportunities and support for students to share their strategies, which often differ. This allows students to take part in class discussions and introduces them to alternative ways to think about the mathematics in the unit.
- In some cases, it may be necessary to read the problems to students or to pair students to facilitate their understanding of the printed materials.
- A list of the materials needed for this unit is in the chart on page xiii.
- Try to follow the recommended pacing chart on page xiii. You can easily spend much more time on this unit than the number of class periods indicated. Bear in mind, however, that many of the topics introduced in this unit will be revisited and covered more thoroughly in other *Mathematics in Context* units.

Resources

For Teachers	For Students
Books and Magazines • *Mathematics Assessment: Myths, Models, Good Questions, and Practical Suggestions*, edited by Jean Kerr Stenmark (Reston, Virginia: The National Council of Teachers of Mathematics, Inc., 1991)	**Videos** Mathsphere Videos • *A Giant Leap* • *A Question of Distribution* (available from Encyclopædia Britannica)

Assessment

Planning Assessment

In keeping with the NCTM Assessment Standards, valid assessment should be based on evidence drawn from several sources. (See the full discussion of assessment philosophies in the *Teacher Resource and Implementation Guide.*) An assessment plan for this unit may draw from the following sources:

- Observations—look, listen, and record observable behavior.
- Interactive Responses—in a teacher-facilitated situation, note how students respond, clarify, revise, and extend their thinking.
- Products—look for the quality of thought evident in student projects, test answers, worksheet solutions, or writings.

These categories are not meant to be mutually exclusive. In fact, observation is a key part in assessing interactive responses and also key to understanding the end results of projects and writings.

Ongoing Assessment Opportunities

- **Problems within Sections**
  To evaluate ongoing progress, *Mathematics in Context* identifies informal assessment opportunities and the goals that these particular problems assess throughout the Teacher Guide. There are also indications as to what you might expect from your students.

- **Section Summary Questions**
  The summary questions at the end of each section are vehicles for informal assessment (see Teacher Guide pages 22, 46, 62, 74, and 86).

End-of-Unit Assessment Opportunities

In the back of this Teacher Guide, there are four assessments that can be given at the end of the unit. Together, they will take up approximately two 45-minute class sessions. For a more detailed description of the assessment activities, see the Assessment Overview (Teacher Guide pages 90 and 91).

You may also wish to design your own culminating project or let students create one that will tell you what they consider important in the unit. For more assessment ideas, refer to the chart on pages xvi and xvii.

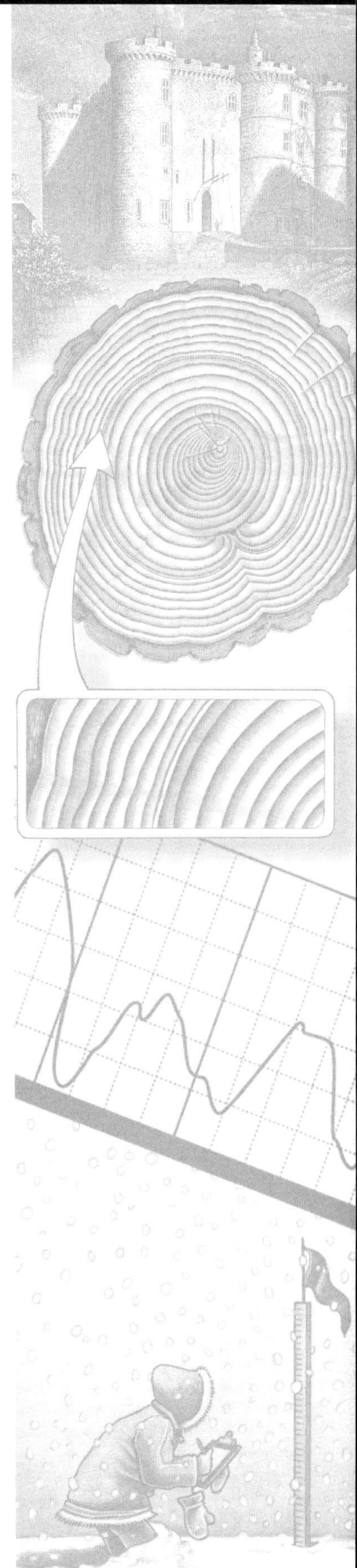

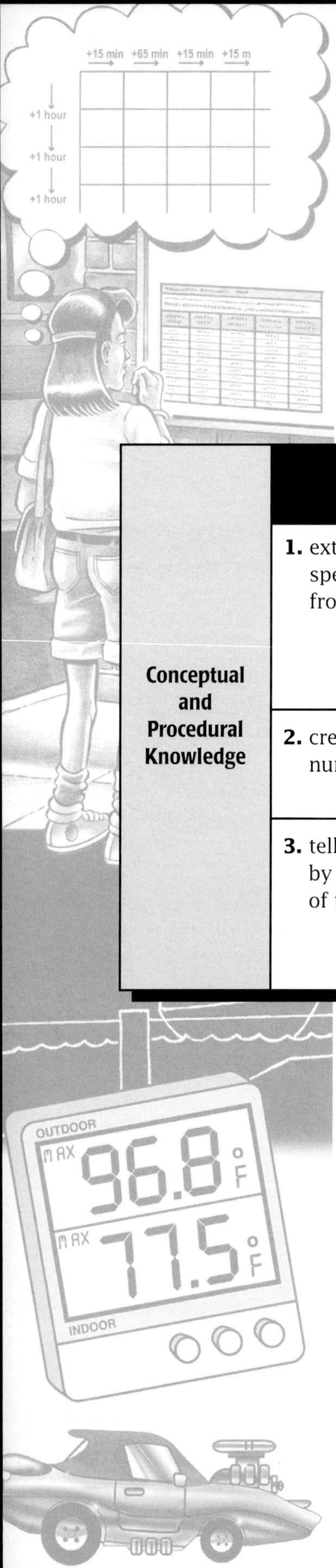

Goals and Assessment

In the *Mathematics in Context* curriculum, unit goals, categorized according to cognitive procedures, relate to the strand goals and to the NCTM Curriculum and Evaluation Standards. Additional information about these goals is found in the *Teacher Resource and Implementation Guide.* The *Mathematics in Context* curriculum is designed to help students develop their abilities so that they can perform with understanding in each of the categories listed below. It is important to note that the attainment of goals in one category is not a prerequisite to attaining those in another category. In fact, students should progress simultaneously toward several goals in different categories.

	Goal	Ongoing Assessment Opportunities	End-of-Unit Assessment Opportunities
Conceptual and Procedural Knowledge	**1.** extract local, point-specific information from line graphs	**Section A** p. 20, #21d p. 24, #23 **Section B** p. 34, #14c, d **Section C** p. 52, #1 **Section D** p. 72, #7, #8 **Section E** p. 88, #9	Presidential Campaign, p. 127 Trends for Women in the Workforce, p. 131 Barcelona Revisited, p. 132
	2. create line graphs from numeric data	**Section B** p. 34, #14a **Section D** p. 72, #6 **Section E** p. 84, #6	Barcelona Revisited, p. 132 Temperature Graphs Revisited, p. 134
	3. tell the story depicted by a line graph in terms of the context	**Section A** p. 10, #8 p. 20, #21d **Section B** p. 44, #20 p. 46, #22 **Section E** p. 88, #10	Presidential Campaign, p. 127 Trends for Women in the Workforce, p. 131 Barcelona Revisited, p. 132

	Goal	Ongoing Assessment Opportunities		End-of-Unit Assessment Opportunities
Reasoning, Communicating, Thinking, and Making Connections	**4.** interpolate data points to fill in line graphs when appropriate	**Section B** **Section E**	p. 34, #14a p. 84, #6	Barcelona Revisited, p. 132 Temperature Graphs Revisited, p. 134
	5. reason about global information such as regularity, maximum/ minimum values, slope, and variance as shown in line graphs	**Section A** **Section B** **Section C** **Section D** **Section E**	p. 24, #23 p. 34, #14b, d, e p. 60, #6 p. 72, #7, #8 p. 80, #2	Presidential Campaign, p. 127 Trends for Women in the Workforce, p. 131
	6. create line graphs from verbal descriptions	**Section D**	p. 74, #9	Temperature Graphs Revisited, p. 134

	Goal	Ongoing Assessment Opportunities		End-of-Unit Assessment Opportunities
Modeling, Nonroutine Problem-Solving, Critically Analyzing, and Generalizing	**7.** understand limitations of line graphs	**Section B** **Section C**	p. 40, #19 p. 56, #3d p. 64, #10c	Barcelona Revisited, p. 132
	8. recognize the importance of scale in creating line graphs	**Section C** **Section D**	p. 56, #3d p. 64, #10c p. 74, #9	Barcelona Revisited, p. 132
	9. recognize the power of graphs for representing information and solving problems	**Section E**	p. 80, #2 p. 88, #9	Trends for Women in the Workforce, p. 131 Temperature Graphs Revisited, p. 134

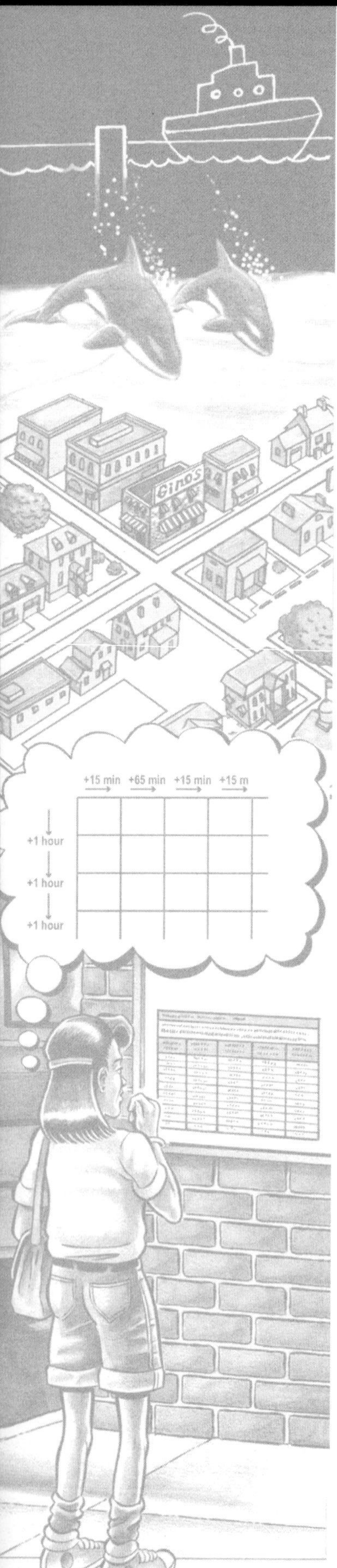

More about Assessment

Scoring and Analyzing Assessment Responses

Students may respond to assessment questions with various levels of mathematical sophistication and elaboration. Each student's response should be considered for the mathematics that it shows, and not judged on whether or not it includes an expected response. Responses to some of the assessment questions may be viewed as either correct or incorrect, but many answers will need flexible judgment by the teacher. Descriptive judgments related to specific goals and partial credit often provide more helpful feedback than percent scores.

Openly communicate your expectations to all students, and report achievement and progress for each student relative to those expectations. When scoring students' responses, try to think about how they are progressing toward the goals of the unit and the strand.

Student Portfolios

Generally, a portfolio is a collection of student-selected pieces that is representative of a student's work. A portfolio may include evaluative comments by you or by the student. See the *Teacher Resource and Implementation Guide* for more ideas on portfolio focus and use.

A comprehensive discussion about the contents, management, and evaluation of portfolios can be found in *Mathematics Assessment: Myths, Models, Good Questions, and Practical Suggestions,* pp. 35–48.

Student Self-Evaluation

Self-evaluation encourages students to reflect on their progress in learning mathematical concepts, their developing abilities to use mathematics, and their dispositions toward mathematics. The following examples illustrate ways to incorporate student self-evaluations as one component of your assessment plan.

- Ask students to comment, in writing, on each piece they have chosen for their portfolios and on the progress they see in the pieces overall.
- Give a writing assignment entitled "What I Know Now about [a math concept] and What I Think about It." This will give you information about each student's disposition toward mathematics as well as his or her knowledge.
- Interview individuals or small groups to elicit what they have learned, what they think is important, and why.

Suggestions for self-inventories can be found in *Mathematics Assessment: Myths, Models, Good Questions, and Practical Suggestions,* pp. 55–58.

Summary Discussion

Discuss specific lessons and activities in the unit—what the student learned from them and what the activities have in common. This can be done in whole-class discussion, in small groups, or in personal interviews.

Connections across the *Mathematics in Context* Curriculum

Tracking Graphs is one of four algebra strand units at the grade 6/7 level. The map below shows the complete *Mathematics in Context* curriculum for grade 6/7. This will indicate where the unit fits in the algebra strand and in the overall picture.

A detailed description of the units, the strands, and the connections in the *Mathematics in Context* curriculum can be found in the *Teacher Resource and Implementation Guide.*

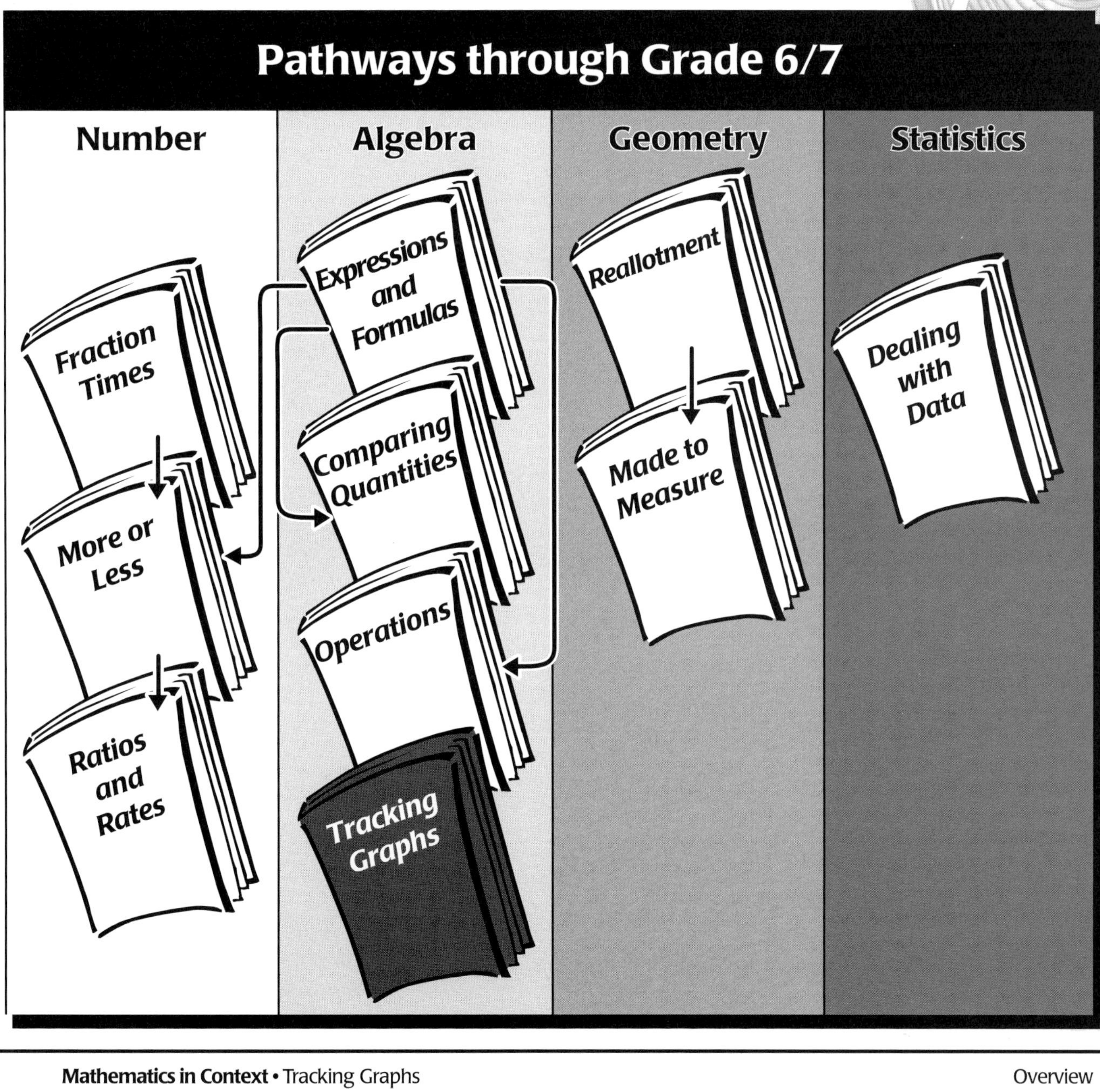

Pathways through the Algebra Strand
(Arrows indicate prerequisite units.)

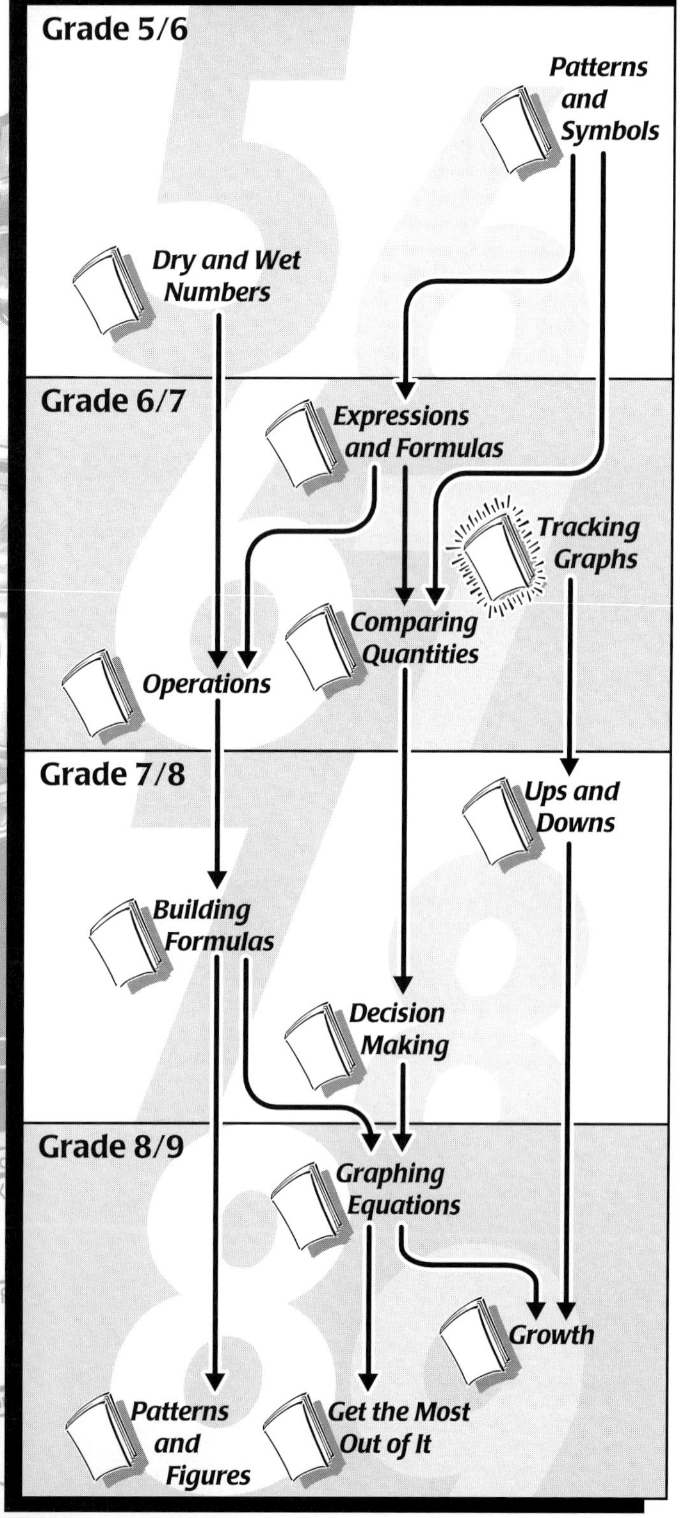

Connections within the Algebra Strand

On the left is a map of the algebra strand; this unit, *Tracking Graphs,* is highlighted.

In grade 6/7, the algebra units build on the concepts introduced in grade 5/6. *Operations* follows the grade 5/6 unit *Dry and Wet Numbers* and develops arithmetic operations with integers. *Expressions and Formulas* and *Comparing Quantities* build on the grade 5/6 unit *Patterns and Symbols,* developing single-variable equations and systems of equations, respectively. *Tracking Graphs* extends the algebra strand skills to include reading, interpreting, and creating graphs. These skills are reviewed and extended in the grade 7/8 unit *Ups and Downs* and in the grade 8/9 unit *Graphing Equations.*

The Algebra Strand

Grade 5/6

Patterns and Symbols
Using symbols to represent patterns, creating equivalent patterns, and making generalizations from them to find a rule.

Dry and Wet Numbers
Developing the concept of negative numbers, adding with positive and negative numbers, interpreting a drawing using a scale, and making a scale line with positive and negative numbers.

Grade 6/7

Expressions and Formulas
Describing series of calculations using operation strings, generalizing to find a formula, and representing relationships found in tables.

Tracking Graphs
Producing line graphs, reading data from graphs, and learning to look at a graph's essential features rather than details.

Comparing Quantities
Organizing and translating information from stories using symbols, charts, tables, and equations.

Operations
Multiplying, dividing, adding, and subtracting positive and negative numbers and plotting ordered pairs in all four quadrants of the coordinate plane.

Grade 7/8

Building Formulas
Relating tables to algebraic descriptions, applying recursive and direct-rule formulas, relating expressions in equivalent forms, and using squares and square roots to solve area problems.

Ups and Downs
Describing an increasing or decreasing function from a table or graph, determining whether or not a growth function is linear, and recognizing a periodic function.

Decision Making
Representing data with graphs, working with the notion of constraints and with graphing inequalities, and graphing discrete functions.

Powers of Ten
Investigating simple laws for calculating with powers of 10 and investigating very large and very small numbers. (*Powers of Ten* is also in the number strand.)

Grade 8/9

Graphing Equations
Graphing points and lines in the coordinate plane, solving single-variable linear equations, using inequalities to describe restricted regions on a graph, and learning the structure of the equation of a line.

Patterns and Figures
Recognizing regularity in patterns; expressing generalities; and exploring progressions, rectangular and triangular numbers, and Pascal's triangle.

Growth
Investigating linear, quadratic, cubic, and exponential functions; using recursive and direct formulas; and describing growth with graphs.

Get the Most Out of It
Solving word problems with two unknowns; solving systems of equations; graphing lines, inequalities, and hyperbolas; and working with curved feasible regions.

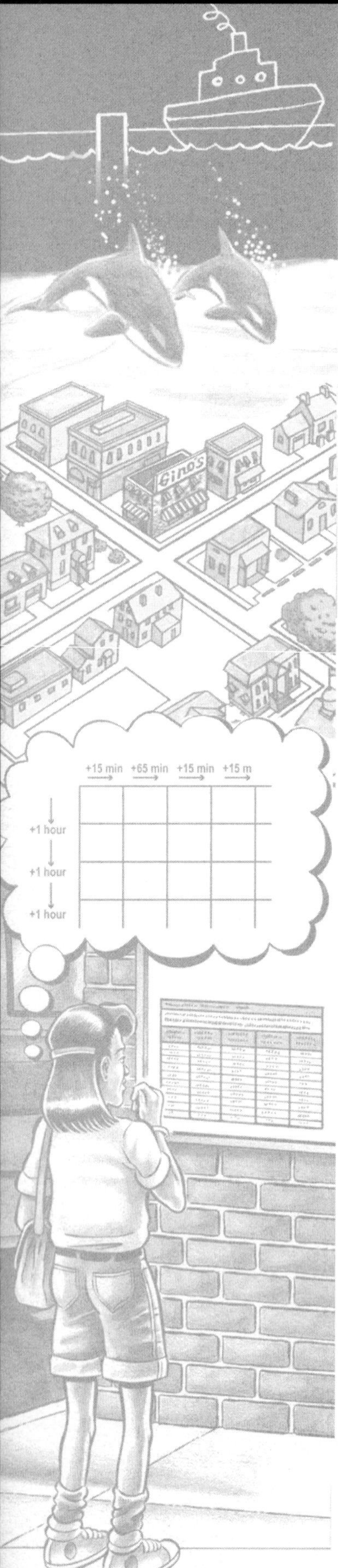

Connections with Other *Mathematics in Context* Units

Tracking Graphs is one of four units in the algebra strand. This unit sets the framework for work in the grade 7/8 and grade 8/9 units dealing with graphs in general and especially those units dealing with graphs that are functions of time.

Graphs as representations of information play an important role in the statistics strand, beginning with the unit *Picturing Numbers,* in which temperature graphs are investigated. Different graphical representations are used and investigated in this strand, as the units *Dealing with Data* and *Insights into Data* show. Connections are made with the geometry strand: the concept of steepness or slope is further investigated in the units *Looking at an Angle* and *Going the Distance.*

The following mathematical topics included in the unit *Tracking Graphs* are introduced or further developed in other *Mathematics in Context* units.

Prerequisite Topics

Topic	Unit	Grade
scale line	*Grasping Sizes**	5/6
representing information graphically	*Picturing Numbers***	5/6

Topics Revisited in Other Units

Topic	Unit	Grade
line graphs	*Ups and Downs*	7/8
	Get the Most Out of It	8/9
	Growth	8/9
slope (as steepness)	*Looking at an Angle****	7/8
	Graphing Equations	8/9
	*Going the Distance****	8/9
representing information graphically	*Dealing with Data***	6/7
	*Insights into Data***	8/9

* This unit in the number strand also helps students make connections to ideas about algebra.
** These units in the statistics strand also help students make connections to ideas about algebra.
*** These units in the geometry strand also help students make connections to ideas about algebra.

Student Materials and Teaching Notes

Student Book
Table of Contents

Dear Student,

Did you ever wonder about the tides of the ocean? Did you ever think about how much the temperature changes in one year? Did you ever investigate the way people run races? In this unit, *Tracking Graphs*, you will use graphs to investigate tides, temperatures, and races. Graphs will help you better understand many interesting things.

In this unit, you will be given information about different situations, and you will use graphs to gain further understanding.

Once you know how to make and interpret graphs, you will find it easier to draw conclusions and explain various situations. This will make you a very powerful mathematical thinker.

Sincerely,

The Mathematics in Context Development Team

Work Students Do

Students study a graph of the men's 100-meter final race in Barcelona during the 1992 Summer Olympics. They learn to identify the runners' positions at various moments—called "snapshots"—in the race. Students investigate a 2,000-meter race between two middle school athletes, estimate and graph the results of the race, and write an article. Finally, students study other graphs of races and write articles.

Goals

Students will:

- extract local, point-specific information from line graphs;
- create line graphs from numeric data;*
- tell the story depicted by a line graph in terms of the context;
- interpolate data points to fill in line graphs when appropriate;*
- reason about global information such as regularity, maximum/minimum values, slope, and variance as shown in line graphs;
- recognize the power of graphs for representing information and solving problems.*

** These goals are introduced in this section and are assessed in other sections of the unit.*

Pacing

- approximately three 45-minute class sessions

Vocabulary

- graph mask
- line graph

About the Mathematics

The main goal of this section is for students to be able to read, create, and interpret line graphs both globally and locally, at a point-specific level. In this unit, line graphs are presented initially as a series of snapshots of a top view of a race. Dots on the graph show the positions of the runners at specific moments in time. Students make line graphs for the first time by putting together snapshots for a race between two runners from different schools. A graph mask that allows only a single point in time to be shown illustrates to students the connection between the individual points and the continuous nature of the line graph. The section ends with looking at graphs globally, first by looking at slope and then by looking at the intersection, or crossing, of lines.

Materials

- Letter to the Family, page 112 of the Teacher Guide (one per student)
- Student Activity Sheets 1–4, pages 113–116 of the Teacher Guide (one of each per student)
- blue and red markers, pages 15, 17, and 19 of the Teacher Guide (one of each per student or per group)
- scissors, page 17 of the Teacher Guide (one pair per student)
- transparencies of Student Activity Sheet 3, page 17 of the Teacher Guide, optional (one per pair or per group)
- overhead projector, page 17 of the Teacher Guide, optional (one per class)

Planning Instruction

You may introduce this section by discussing data about local track races from a newspaper. You might also discuss world record holders for the 100-meter dash.

Students may work in pairs or small groups on problems 1–7, 9–20, and 22. Problems 8, 21a–c, and 23 can be worked on either individually or in pairs. Problem 21d is ideal for individual student work.

There are no optional problems in this section.

Homework

Problems 8 (page 10 of the Teacher Guide), 18 (page 18 of the Teacher Guide), 22 (page 22 of the Teacher Guide), and 23 (page 24 of the Teacher Guide) may be assigned as homework. The Writing Opportunity (page 11 of the Teacher Guide), and the Extension activities (pages 11 and 21 of the Teacher Guide) can also be assigned as homework. After students complete Section A, you may assign appropriate activities from the Try This! section, located on pages 39–42 of the *Tracking Graphs* Student Book. The Try This! activities reinforce the key mathematical concepts introduced in this section.

Planning Assessment

- Problem 8 can be used to informally assess students' ability to tell the story depicted by a line graph in terms of the context.
- Problem 21d can be used to informally assess students' ability to extract local, point-specific information from line graphs and to tell the story depicted by a line graph in terms of the context.
- Problem 23 can be used to informally assess students' ability to extract local, point-specific information from line graphs and to reason about global information such as regularity, maximum/minimum values, slope, and variance as shown in line graphs.

A. GRAPHING A RACE

BARCELONA 1992

In 1992, the Summer Olympics took place in Barcelona, Spain. One of the most publicized events was the men's 100-meter dash. Eight runners qualified for the final.

The following day, many articles about the final race were in newspapers. One of the articles was illustrated with the graph below. This graph is enlarged on **Student Activity Sheet 1.** Use the graph to help you answer Problems **1–8.**

1. a. Which countries were represented in the race?

b. Who won the race, and in what time did he win it?

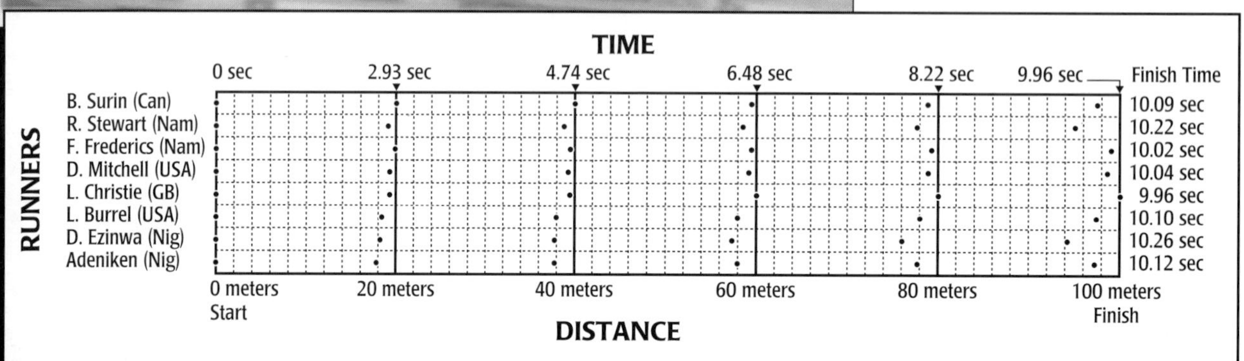

Solutions and Samples
of student work

1. **a.** Canada, Namibia, the United States, Great Britain, and Nigeria

   b. L. Christie from Great Britain, in 9.96 seconds

Hints and Comments

Materials Student Activity Sheet 1 (one per student)

Overview Students study a graph (diagram) that represents the 100-meter men's final at the 1992 Olympics.

About the Mathematics The graph shows each runner's position at different moments—at 20-meter intervals—in the race. It shows the relationship between time and distance for each runner for six points or times in the race. Each time or "snapshot" is determined by the runner who reaches the 20-meter interval first. Reading from left to right, students should be able to understand the progress of the race. This problem should develop students' ability to read, create, and interpret information from line graphs.

Planning Students may work in pairs or in small groups on problem **1.**

Comments about the Problems

1. Each time listed across the top of the graph represents when the first runner reached that position (that 20-meter interval).

   b. The finish times are to the right of the graph. The eight dots just to the left of these times represent a snapshot of the race at the moment Christie passed the finish line.

2. **a.** Who were the U.S. runners?

   b. Did they win any medals?

Now that you know who won, look at what happened during the race. The horizontal rows on the graph represent the lanes of the runners. At the left end, you see the finalists in starting position at the moment of the starting shot.

The second column of dots is a "snapshot" of the race 2.93 seconds after the start.

3. **a.** Who was in the lead at this moment?

   b. Who was in second position, and who was third?

The next snapshot is 4.74 seconds into the race. Surin, the Canadian, was in the lead.

4. Which stretch did Surin run faster, the first 20 meters of the race or the second 20 meters? Explain.

Solutions and Samples
of student work

2. **a.** Mitchell and Burrel

   b. Mitchell won the bronze medal for third place.

3. **a.** Surin

   b. In second position was Frederics, and in third were Mitchell and Christie.

4. Surin ran the second 20 meters faster than he ran the first 20 meters. Strategies will vary. Sample strategies:

   Strategy 1

   Surin ran the 40 meters in 4.74 seconds. Suppose he ran both stretches with the same speed: then he would have run each in 4.74 ÷ 2 = 2.37 seconds. But he reached the 20-meter line in 2.93 seconds, so he ran the first 20 meters slower.

   Strategy 2

   Suppose Surin ran the second 20 meters with the same speed as he ran the first 20 meters. Then he would have required 2.93 + 2.93 = 5.86 seconds. The actual time was 4.74 seconds, so he must have run faster in the second 20 meters.

   Strategy 3

   Surin ran the first 20 meters in 2.93 seconds and the second 20 meters in 4.74 – 2.93 = 1.81 seconds. He ran the second 20 meters faster than he ran the first 20 meters.

   Explanations will vary. Some students may say that at the beginning of a race, it takes time for the runners to get their speed up.

Hints and Comments

Materials Student Activity Sheet 1 (one per student)

Overview Students interpret the 20-meter and 40-meter snapshots on the race graph on Student Activity Sheet 1. They use time and distance to make comparisons in terms of speed.

About the Mathematics Students extract information from the graph, make conclusions based on this information, and prove the validity of these conclusions using the graph. This involves comparing decimal numbers written to the hundredths.

Planning Students may continue to work in pairs or in small groups on problems **2–4.** When students have finished problem **4,** discuss their solutions and strategies.

Comments about the Problems

2. b. Students may use different strategies to solve this problem. They may compare the positions of the dots in the last snapshot of the diagram, or they may compare the finishing times given to the right of the diagram. For this last strategy, students must be able to compare decimal numbers.

3. Students should understand that the time on top of the 20-meter line is the time Surin needed for the first 20 meters; the graph does not show when the other runners passed the 20-meter line. This snapshot only shows where they are at that moment.

4. If students answer this question with the generalization that sprinters gain momentum throughout the race, encourage them to support their statement with calculations.

The next snapshot is 6.48 seconds into the race.

5. **a.** Who was the first to reach the 60-meter mark?

b. What was his time at that point in the race?

c. Who do you think was the last one to reach the 60-meter mark? Can you be sure?

6. **a.** In what place did Surin finish the race?

b. How far behind was Surin when the winner was at the finish line?

7. **a.** How far apart were the fastest and slowest runners at 2.93 seconds?

b. How far apart were the fastest and slowest runners at 9.96 seconds?

c. Describe how the distance between the fastest and slowest runners changed over the course of the race.

8. Write a paragraph comparing the race Mitchell ran with the race Christie ran.

Solutions and Samples
of student work

5. a. Christie

b. 6.48 seconds

c. You cannot be certain, but it was probably Ezinwa (Nigeria). Explanations will vary. Sample explanation:

Look before the 6.48 snapshot; Ezinwa is the farthest from the 60-meter mark and would not likely pass anyone else before the 60-meter mark.

6. a. fourth

b. Answers will vary. Some students may say that he was about 2.5 meters, or between 2 and 3 meters, behind. Other students may say that he was $10.09 - 9.96 = 0.13$ seconds behind.

7. a. a little more than 2 meters

b. almost 6 meters, or between 5 and 6 meters

c. Descriptions will vary. Some students may say that, for the first 60 meters, the distance between the fastest and slowest runners was relatively consistent (a little more than 2 meters), but it increased over the last 40 meters.

8. Answers will vary. Sample response:

Mitchell ran a very even race in that he was always within 1 meter of the lead. Christie ran even with Mitchell, but then was able to speed up enough to take the lead after 60 meters. Christie's lead over Mitchell at the finish line was about 1.5 meters.

Hints and Comments

Materials Student Activity Sheet 1 (one per student)

Overview Using the race graph on Student Activity Sheet 1, students compare the positions of the runners at different times and estimate the distances between them. Students finish analyzing the 100-meter race by extracting all the information they can about the entire race as run by two of the runners.

Planning Students may work in pairs or in small groups on problems **5–7.** Students may work individually or in pairs on problem **8.** Discuss students' answers for problems **7** and **8.** Problem **8** may be used as informal assessment and assigned as homework.

Comments about the Problems

6. a. Students can use one of two strategies to get an answer: some students will compare the locations of the dots in the graph, while others will compare the times.

b. Some students might give an answer in terms of time rather than distance. This is acceptable. Note that the distance between two dotted vertical lines is 2 meters, so halfway between two lines is 1 meter, and $\frac{1}{4}$ of the distance between two lines is $\frac{1}{2}$ meter.

8. Informal Assessment This problem assesses students' ability to tell the story depicted by a line graph in terms of the context. It can also be assigned as homework.

Writing Opportunity You may ask students to write their answers to problem **8** in their journals.

Extension You may discuss the following questions as a class or assign this Extension as homework.

- What would the graph look like if a runner fell?
- Suppose you had a similar graph for the 200-meter dash. How do you think this graph would compare to the 100-meter graph?

THE SCHOOL RACE

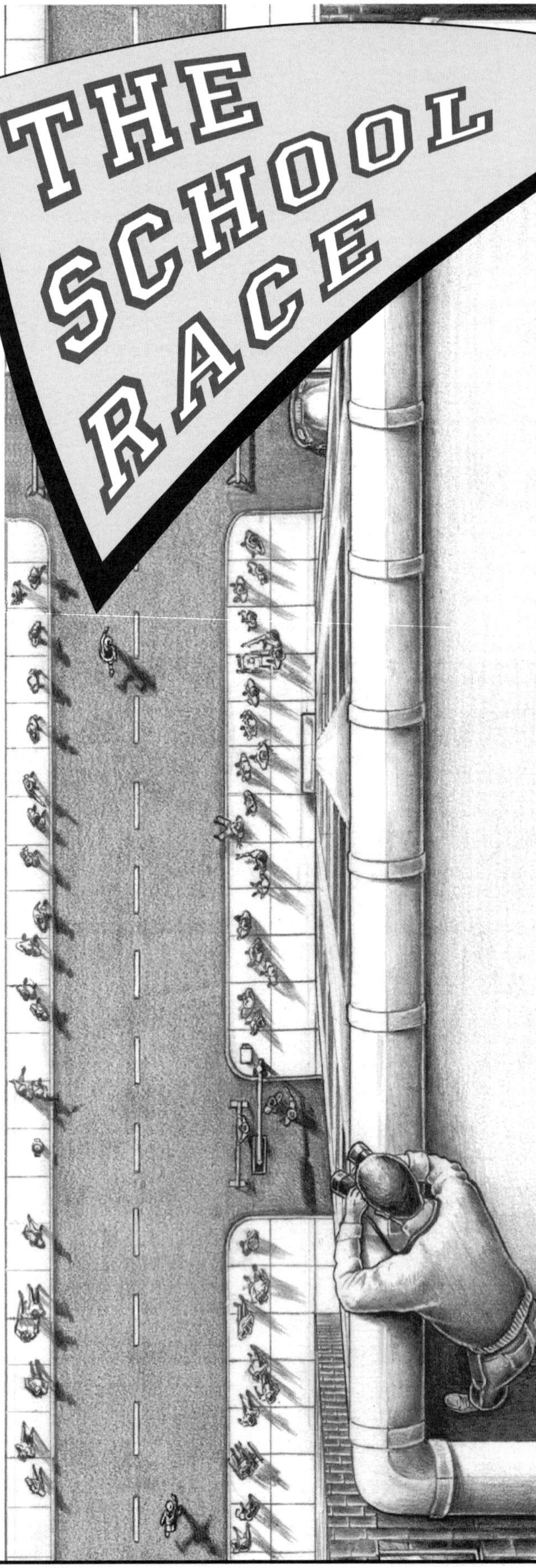

The two middle schools in Middleton—Thomas Jefferson and Martin Luther King, Jr.—are known for their outstanding achievements in track-and-field. Every year, there is an all-city meet to determine which school has the best athletes. One of the events at the all-city meet is the girls' 2,000-meter race. Each school selects one runner for this race. The 2-kilometer course is on a straight road through the middle of town.

You have been asked to write an article about the race for tomorrow's newspaper. In order to have a good view of the whole race, you stand on the roof of a nearby tall building.

Maria, representing Jefferson, is dressed in blue. Jackie, representing King, is in red.

After three minutes, Maria passes the 600-meter point, and she has a considerable lead of about 50 meters.

9. Estimate the winning time for this race.

9. Estimates will vary. Some students may estimate the winning time of the race to be about 10 minutes. Strategies will vary. Sample strategies:

If Maria runs 600 meters in 3 minutes, this equals 200 meters a minute, so 2,000 meters would take about 10 minutes.

Using a ratio table:

÷ 3 × 10

Time (minutes)	3	1	10
Distance (meters)	600	200	2,000

Using a double number line:

Overview Students continue to study the context of track races by investigating a 2,000-meter race between two athletes from different middle schools. They use the lead runner's time at the 600-meter point to estimate the time of the winner.

About the Mathematics In the Olympic 100-meter dash context, students saw that the speed of a runner is not the same at every moment of the race. In other words, the ratio between time and distance is not constant. In this context, students estimate the winning time in a 2,000-meter race using the information that the lead runner at 600 meters has a time of 3 minutes. Here they must make the assumption that the ratio between time and distance stays the same.

Planning Students may work in pairs or in small groups on problem **9.** Discuss students' estimates of the distance between the runners shown in the top view by using the squares of the sidewalk. If the squares are about 1 meter on a side, 19 squares would give a distance of nearly 19 meters.

Comments about the Problems

9. Ask for students' reasoning. Students should use the given distance and time (600 meters in 3 minutes) rather than their own experience or knowledge of the time in which a middle school student can run 2,000 meters. If students have difficulties organizing their calculations, you may encourage them to use a double number line or a ratio table.

Use **Student Activity Sheet 2** to answer problems **10–13.**

10. The diagram on the right shows the course as seen from above. In figure 1 on **Student Activity Sheet 2,** mark the positions of the runners after 3 minutes. Put colored dots inside the bar: a blue dot for Maria and a red dot for Jackie.

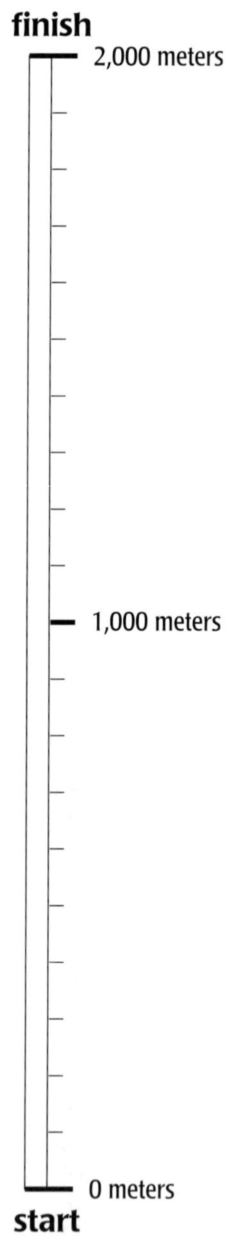

After 5 minutes, the runners are halfway through the race. Because she pushed so hard to have a good start, Maria is tired and has slowed down a bit. Jackie uses this opportunity to pass Maria, but Maria is running just behind her.

11. In figure 2, show the situation in the race after 5 minutes.

After 7.5 minutes, Jackie has extended her lead to 50 meters and passes the 1,500-meter point.

12. In figure 3, show the situation in the race after 7.5 minutes.

After 9 minutes, Maria has a burst of speed. With an extreme effort, she manages to catch up with Jackie. By this time, both runners are 200 meters from the finish line.

Near the finish, Jackie notices Maria running beside her, so she starts her final sprint. Maria, still tired from her effort to catch up, lets Jackie go. Jackie wins the race in 10 minutes.

13. In figures 4 and 5, draw the situations after 9 and 10 minutes.

Solutions and Samples
of student work

10–13.

Figure 1:

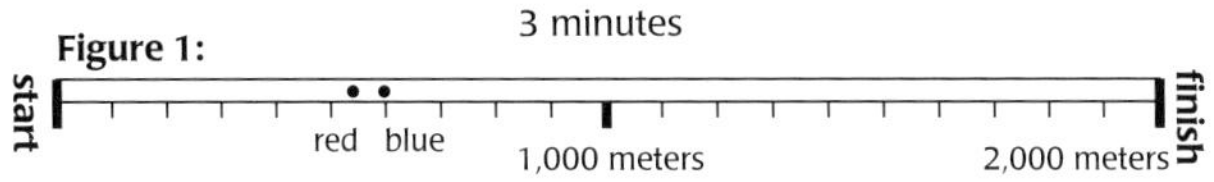

Figure 2:

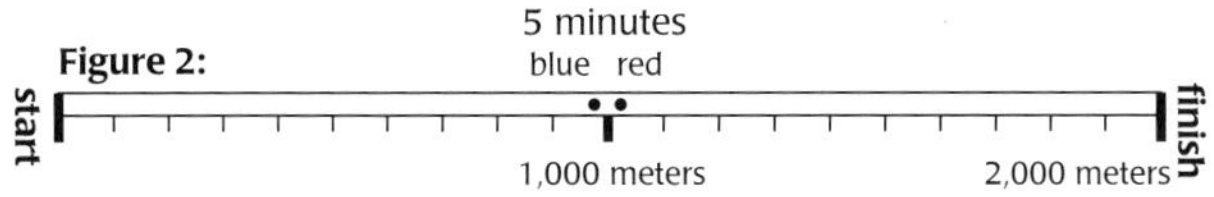

Figure 3:

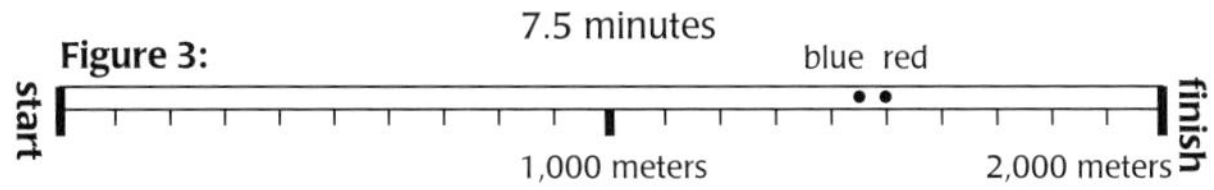

Figure 4:

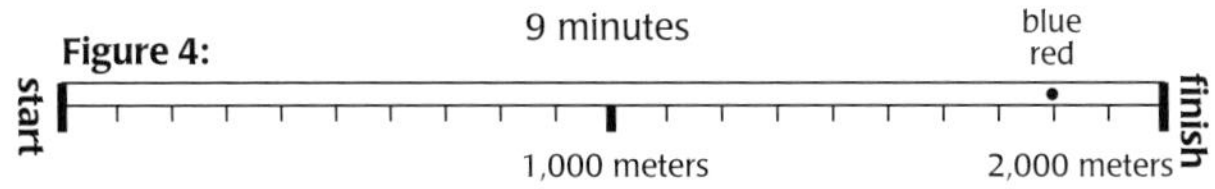

Figure 5:

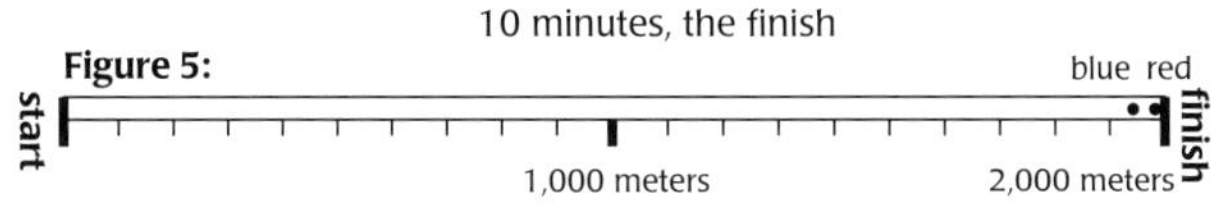

Hints and Comments

Materials Student Activity Sheet 2 (one per student); blue and red markers (one of each per student or per group)

Overview Students create "snapshots" of a race by putting dots on a bar to indicate the locations of two runners at given time intervals. They use different-colored bars for different intervals.

Planning Students may work on problems **10–13** in pairs or small groups. You may want students to compare their drawings with those of other members of the group.

Comments about the Problems

10–13. Make sure that students are putting the dots inside the bar and not beside it.

11. Let students decide where they want to position the dots. Students should realize that if the runners are close in the race, they will look even closer as dots on the scale. If students are having difficulty, you might remind them that the graph is viewed from above. Two separated dots would represent runners who are at least 10 meters apart.

Students may suggest coloring one dot half red and half blue. This situation previews students' work in intersecting two graphs. You may want to discuss this when students create the graphs in the problems on the next page.

Your editor wants you to make a graph to go with your newspaper article that will show readers, at a single glance, how the race progressed. You can make this graph by combining the information from all five of your pictures on **Student Activity Sheet 2.**

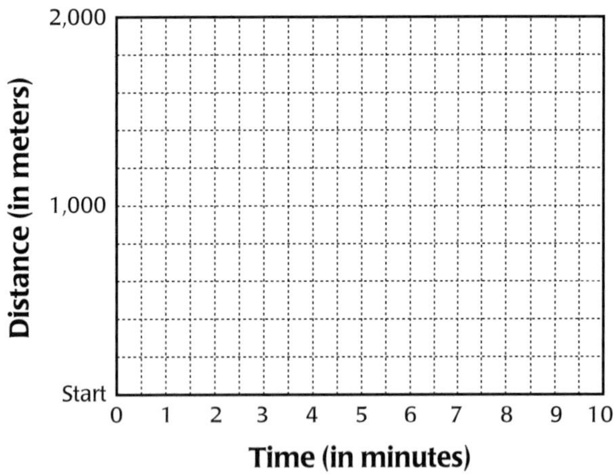

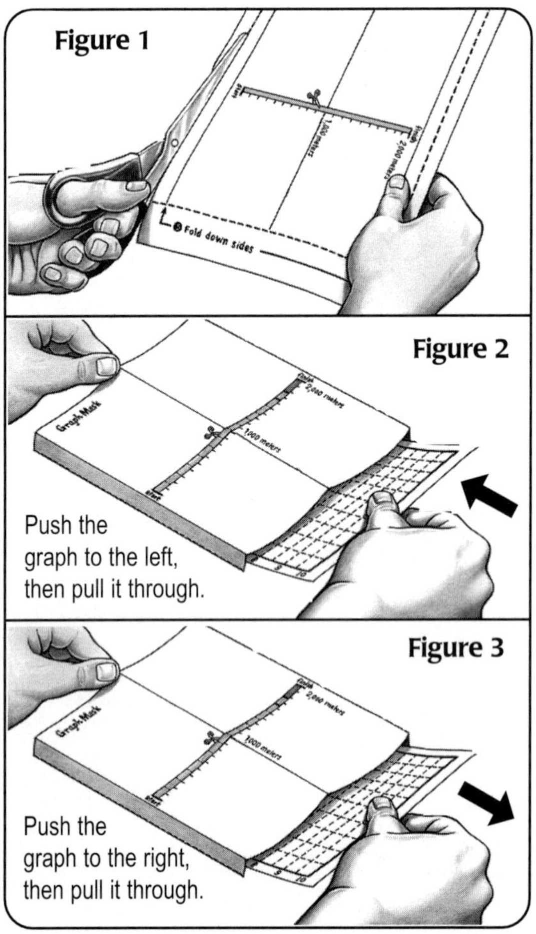

14. Use **Student Activity Sheet 3** to create your graph by following these steps:

- Still using blue for Maria and red for Jackie, put all five situations from **Student Activity Sheet 2** in the proper places on this single graph.
- Add dots to this graph to show the positions of the runners at the start of the race.
- Add dots to this graph to show what the positions of the runners might have been at every half minute of the race.
- Connect all of the blue dots. Connect all of the red dots.
- Cut out your graph along the heavy dashed lines as shown.

15. What does each curve show?

16. Compare your graph with other students' graphs. Describe some differences.

Using the graph mask on **Student Activity Sheet 4** along with the graph you just cut out will help you visualize the race.

- Follow the instructions for cutting out and folding the graph mask. (See figure 1.)
- Position the graph under the mask. Keeping the mask still, push the graph to the left. Then pull it the rest of the way through to see the progress of the race. (See figure 2.)

17. What would you see if you moved the graph from the left to the right? (See figure 3.)

Solutions and Samples
of student work

14.

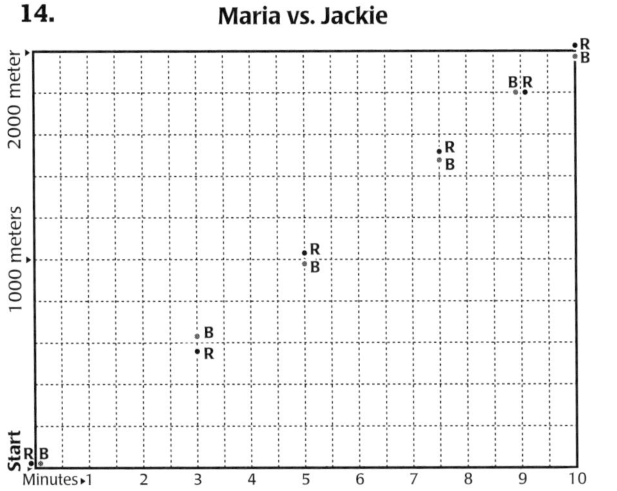

Line graphs will vary. Sample student graph:

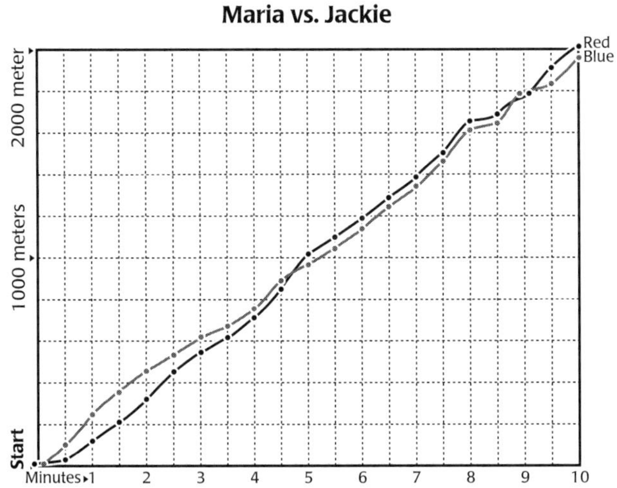

15. Descriptions will vary. Sample response:

The lines show the positions of the runners at any time during the race.

16. Descriptions will vary. Some lines will be further apart or closer together. Some lines will be smoother than others.

17. the race in reverse

Hints and Comments

Materials Student Activity Sheets 2–4 (one of each per student); blue and red markers (one of each per student or per group); scissors (one pair per student); transparencies of Student Activity Sheet 3, optional (one per pair or per group); overhead projector, optional (one per class)

Overview Students create a line graph to show the progress of two runners in a race.

About the Mathematics The five bars students created for the problems on Student Book page 5 show the positions of the runners at five different moments. The two curves on this line graph will show the position of each runner at every moment of the race. Each dot on the graph represents a time and the corresponding distance of the runner. Note that the time can be read on the horizontal axis and the distance on the vertical axis. The graph mask will help students visualize the race. This will also further develop their reading of graphs. Notice that this graph is different than the one discussed at the beginning of the section. Distance is now along the vertical axis and time is along the horizontal axis. This makes it easy for students to see who is in the lead (the highest dot) at any moment.

Planning You may want to make transparencies of Student Activity Sheet 3 for each pair or small group of students. Then students can easily show their results to the class. Students may work on problems **14–17** in pairs or in small groups. Discuss students' solutions as a class.

Comments about the Problems

14. This is a critical problem because students are expected to be able to make a line graph for the first time. Observe students as they position their dots on the grid paper. They should see that the distance between two lines represents 200 meters. Graphs will vary slightly from group to group.

There is no one correct graph. Results depend on where a student puts the dots for each half minute. Be sure that all students have the graph starting at zero time, zero distance.

16. This problem will invite students to look at the graphs in a more global way. In discussing differences, be sure to relate the graphs to the real race situation.

18. Write an article for the newspaper describing the race. Write it so the reader will refer to the graph while reading the article.

Last Year's *RACE*

Here is a description of last year's race:

The runners were not able to break away from each other during the first 3 minutes. In that time, they covered 650 meters. Then the runner from King stumbled, and the Jefferson runner took the lead. Six minutes into the race, Jefferson's runner was at 1,300 meters, and her lead was at its maximum (about 150 meters). The runner from King managed to regain her speed but could never catch up.

Results:

First Place—Jefferson
(9 minutes, 30 seconds)

Second Place—King
(10 minutes)

19. Make a graph for last year's race on another copy of **Student Activity Sheet 3.** Work in pairs or groups.

Suppose you were going to judge the graph of another group.

20. a. Why should you check to see that both lines start at the bottom left corner?

b. Write three other things a graph of the race should have. Be sure your own graph fits your criteria.

Solutions and Samples
of student work

18. Articles will vary, but should include the following information:
- the total distance of the race;
- the number of runners;
- who takes the lead at the beginning of the race;
- the time and/or distance at which a runner passes another runner;
- the situation just before the finish; and
- the results of the race and finishing times.

19. Answers may vary.

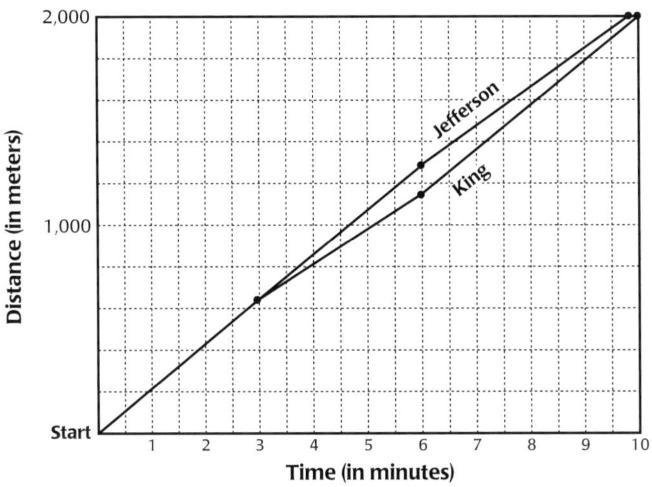

20. a. The bottom left corner represents the start of the race, where time and distance are both zero.

b. Answers will vary. Sample answers:
- The red and blue curves must coincide for the first 650 meters.
- The red curve (King) must be under the blue curve (Jefferson) for the remainder of the race.
- The curves must get closer after 6 minutes.
- The blue curve must end at $9\frac{1}{2}$ minutes, just to the left of the upper right corner of the grid.
- The red curve must end in the upper right corner.

Hints and Comments

Materials Student Activity Sheet 3 (one per student); blue and red markers, optional (one of each per student or per group)

Overview Students write an article that refers to the graph of a race. They then create a new graph of another race between two runners.

About the Mathematics Graphs can be examined in detail, or "locally," and they can be looked at in their entirety, or "globally." The approach depends on the information desired. One has to look locally to extract point-specific data. One has to look globally, at the shape of the graph and how the graph changes, to reason about such things as regularity, an increase, a decrease, constancy, the maximum, and the minimum. To develop students' understanding of graphs, it is important to have students look locally as well as globally from the onset.

Planning Students may work in pairs or groups on problems **18–20.** Problem **18** may be assigned as homework. Discuss students' solutions and strategies for all three problems.

Comments about the Problems

18. Homework This problem may be assigned as homework. Have students share their articles with the class. List with the students all the important facts the graph shows and how to recognize important moments of the race in the graph. This discussion may help students who had problems with this activity get a better idea of what to pay attention to and how global or detailed such stories should be.

19. Students should represent the story by marking the graph with dots. Some will draw the curves right away, while others will need to put intermediate dots for every 30 seconds. You might want to discuss the use of different colors for each curve or the idea of labeling each curve with a name.

This is the graph of a race held in a nearby town.

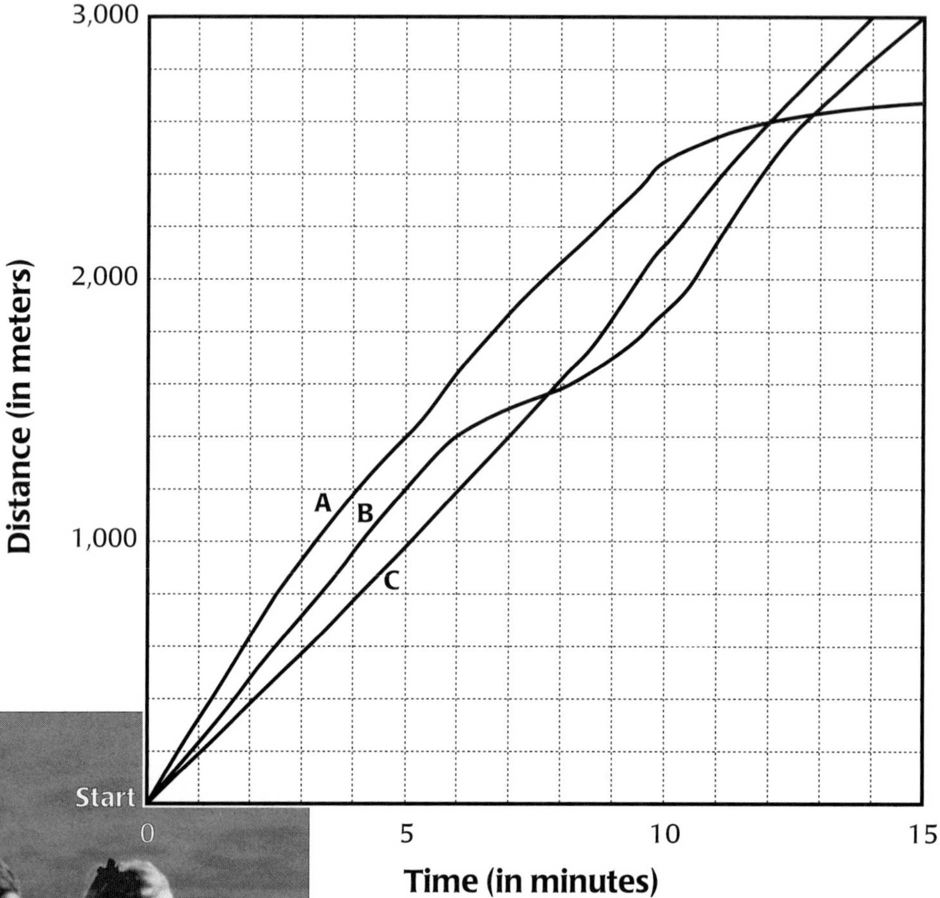

Study the graph shown above. There are some differences between it and the graph of the King vs. Jefferson race that you made for problem **19.**

21. a. What differences can you find?

b. What dramatic thing happened in this race?

c. From start to finish, one of the lines in the graph is almost straight. What does that tell you about the race of this runner?

d. Write a short article about the race. Be sure to include all of the important facts.

Solutions and Samples
of student work

21. a. Answers will vary. Sample responses:

There are three lines on this graph, representing three runners (possibly three schools).

The race is longer (3,000 meters), so it takes more time.

Not all runners finish this race.

b. Answers will vary. Some students may say that runner **A** either dropped out of the race or could not run any more because of an injury. Other students may observe that the line for runner **A** levels off.

c. It tells you that the runner ran the same speed for nearly the whole race.

d. Answers will vary. Sample response:

Runner **A** made a very fast start, and she had a 200-meter lead after 3 minutes. Runner **B** was more than 100 meters ahead of runner **C** at that point. Six minutes into the race, runner **B** slowed down and was passed by runner **C** before they had run about 1,600 meters, almost 8 minutes into the race. Runner **A** had by then increased her lead to over 400 meters. This was her biggest lead. Ten minutes into the race, she twisted her ankle and had to start walking slowly. Runner **C** passed runner **A** at 2,600 meters and never looked back. She won the race in 14 minutes, with runner **B** almost 200 meters behind. Runner **B** finished one minute later.

Hints and Comments

Materials Student Activity Sheet 3, optional (one per student); Student Activity Sheet 4, optional (one per class)

Overview Students extract information from a graph representing a race with three runners.

About the Mathematics Students study the changes and shapes of line graphs. They describe the changes in the graphs in terms of the context of the race. Where two curves intersect, one runner is passing another. Where a curve flattens, a runner is slowing down.

Planning Students may work individually or in pairs on problems **21a–c.** You might have them work individually on problem **21d,** which may be used as informal assessment.

Comments about the Problems

21. Be sure that students are able to read the line graph. After students complete the problem, you may want to demonstrate the use of the graph mask using Student Activity Sheet 4.

b. You might ask students what the line for **A** would suggest if it stayed completely flat after some point, rather than moving up slowly.

c. This might be a difficult question for some students. You might suggest that such students make a guess.

d. Informal Assessment This problem assesses students' ability to extract local, point-specific information from line graphs and to tell the story depicted by a line graph in terms of the context.

Check students' articles for mention of important moments, such as when two runners pass each other. Articles should also give the order in which the runners finish and their finishing times.

Extension You can ask students to draw their own graph of a race with three runners and write a story to describe it. Provide them with a copy of Student Activity Sheet 3, which they can use or modify. You may have students work in pairs or assign this as homework.

Summary

A graph is a picture of a situation. Pictures often tell a story. For example, the graph of the 100-meter race tells the story of that race.

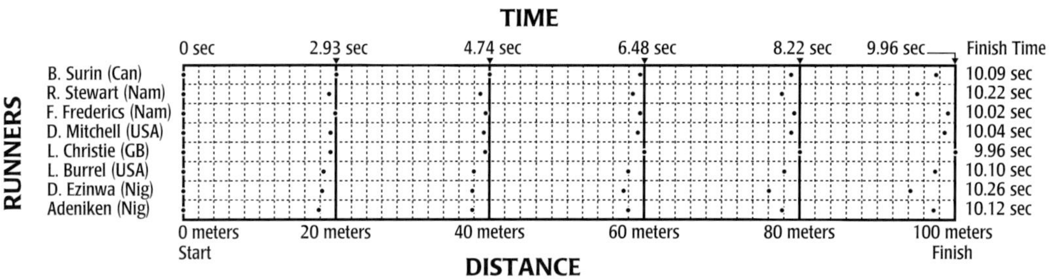

A *line graph* is a special kind of graph. It represents information occurring over time.

Your first line graphs were the red and blue graphs for the school races. Putting lines for two runners on the same graph makes it easy to compare the runners.

Summary Questions

Look again at **Student Activity Sheet 1** to answer the following question.

22. **a.** Who were in the first, second, and third places 8.22 seconds into the race?

b. In what place was Surin at each 20-meter mark?

c. Approximately where and when do you think Christie passed Surin?

Solutions and Samples
of student work

22. a. Christie was first; Frederics was second. Answers for third place will vary. Some students may say that Mitchell was third, based on his third-place finish at the end of the race.

b. 20 meters—first
40 meters—first
60 meters—second or third
80 meters—third or fourth
finish—fourth

c. somewhere between 40 and 60 meters and between 4.74 and 6.48 seconds into the race

Hints and Comments

Materials Student Activity Sheet 1 (one per student)

Overview After reading the Summary, students extract more information from the graph of the Olympic race on Student Activity Sheet 1.

Planning In a discussion with the class about the Summary, you may ask your students what advantages a line graph of the race would have over the graph with the six snapshots. Students may work on problem **22** in pairs or in small groups. This problem may also be assigned as homework.

Comments about the Problems

22. Homework This problem can be assigned as homework.

c. Some students might respond with more exact answers. Look to see that the time is appropriate for the distance. For example, if they say 50 meters, then the time should be about 5.5 seconds. Note that the halfway point between two distances will not always be the halfway point between two times.

Summary Questions, continued

Cecilia and her little sister Jasmine are going to race each other. Cecilia decides to give Jasmine a head start to make the race fair. She waits until Jasmine has reached the 100-meter mark before she starts to run. Here is the graph of their race.

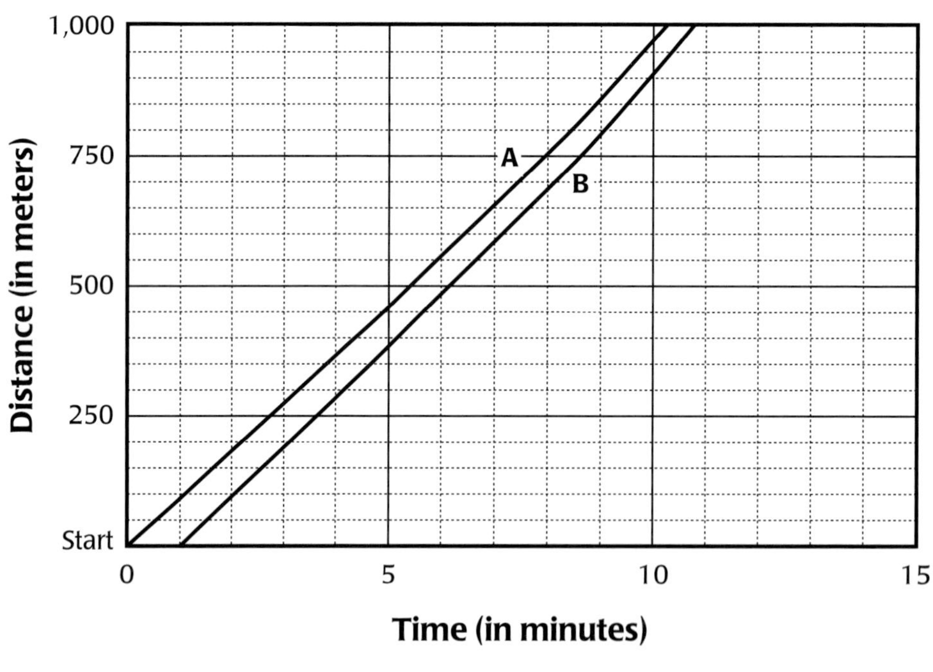

23. **a.** Who is runner A? runner B? How do you know?

b. Who won the race, and how far behind was the other person when the winner crossed the finish line?

c. Who ran faster in this race? How do you know?

Solutions and Samples
of student work

23. a. Jasmine is runner **A.** Explanations will vary. Some students may say that Jasmine is runner **A** because at 1 minute she is at 100 meters, while Cecilia is at 0 meters. Students may also see by looking at the horizontal axis that runner **A** starts first: runner **A** starts when the time starts, and runner **B** starts about one minute later.

b. Jasmine won the race, because she reached the 1,000-meter mark first (at about 10 minutes, 15 seconds). Cecilia was about 75 meters behind.

c. Cecilia ran faster. Explanations will vary. Sample explanations:

- Cecilia started 100 meters behind and finished about 75 meters behind Jasmine, so she gained about 25 meters.
- Jasmine ran the 1,000 meters in just over 10 minutes. Cecilia finished in under 11 minutes, but she started about one minute later.
- The graphs get closer and closer, indicating that the distance between the runners was shrinking and that Cecilia was running faster and gaining.

Hints and Comments

Overview Students extract global and detailed information from a line graph of a race between two runners.

Planning Students may work on problem **23** individually or in pairs. This problem can be assigned as homework and used as informal assessment. After students finish this problem, discuss students' answers and reasoning.

After students complete Section A, you may assign appropriate activities from the Try This! section, located on pages 39–42 of the *Tracking Graphs* Student Book, for homework.

Comments about the Problems

23. Informal Assessment This problem assesses students' ability to extract local, point-specific information from line graphs and reason about global information such as regularity, maximum/minimum values, slope, and variance as shown in line graphs. This problem may also be assigned as homework.

SECTION B. FILLING IN TEMPERATURES

Work Students Do

Students mark temperatures on a graph at different times during a single day. Points in between are filled in using assumptions about temperature changes as well as different weather descriptions. Students then connect the points to show a situation that is continuous. Students explain how they know that a gap in a graph is filled in incorrectly. At the end of the section, students match verbal descriptions to various graphs.

Goals

Students will:

- extract local, point-specific information from line graphs;
- create line graphs from numeric data;
- tell the story depicted by a line graph in terms of the context;
- interpolate data points to fill in line graphs when appropriate;
- reason about global information such as regularity, maximum/minimum values, slope, and variance as shown in line graphs;
- understand limitations of line graphs.

Pacing

- approximately four 45-minute class sessions

About the Mathematics

In this section, line graphs are spoken of in a new way, as the result of a refinement process involving taking more and more measurements. For a temperature graph, the thermometer acts as a mobile scale for plotting dots in a grid to make a graph and for interpreting information from the graph. Interpolating with a graph is presented as finding points that are likely to have occurred between known data points.

Depending on the context, there are restrictions on the rate of increase or decrease in a graph. For many processes that describe a change over time, it is impossible for the graph to have a vertical part, which would indicate more than one data point in a single moment.

Students also look at more global features of the graph: average, slope, and variation. The concept of average, or mean, is taught in the grade 5/6 unit *Picturing Numbers.* In that unit, the compensation strategy is used (read more about this strategy in the About the Mathematics section in the Hints and Comments column on Teacher Guide page 37). Section B ends with a conversion of the vertical scale of a graph from Fahrenheit to Celsius.

Materials

- Student Activity Sheets 5–9, pages 117–121 of the Teacher Guide (one of each per student)
- scissors, page 31 of the Teacher Guide (one pair per student)
- transparency of completed time/temperature chart, pages 31 and 33 of the Teacher Guide, optional (one per class)
- overhead projector, pages 31 and 33 of the Teacher Guide, optional (one per class)
- weather data illustrated in graphs from local newspapers, page 43 of the Teacher Guide, optional (two or three per class)
- graph paper, page 47 of the Teacher Guide (one sheet per student)

Planning Instruction

You may wish to begin the lesson with a discussion of thermometers and how they work, and/or a discussion of the relationship between degrees Celsius and degrees Fahrenheit. On the Fahrenheit scale, water freezes at 32° and boils at 212°; on the Celsius scale, water freezes at 0° and boils at 100°. Changing a line graph of temperatures from degrees Fahrenheit to degrees Celsius would therefore make the graph less steep if the same vertical scale were used. (This is investigated at the end of the section.)

Students can work in pairs or in small groups on problems 1–13, 17–19, and 21–23. You may have them do problem 20 individually or in pairs. They may work on the remaining problems individually.

There are no optional problems in this section.

Homework

Problem 21 (page 46 of the Teacher Guide) may be assigned as homework. In addition, the Extension (page 41 of the Teacher Guide) and the Writing Opportunity (page 43 of the Teacher Guide) may be assigned as homework. After students complete Section B, you may assign appropriate activities from the Try This! section, located on pages 39–42 of the *Tracking Graphs* Student Book. The Try This! activities reinforce the key mathematical concepts introduced in this section.

Planning Assessment

- Problem 14a can be used to informally assess students' ability to create line graphs from numeric data, and to interpolate data points to fill in line graphs when appropriate.
- Problems 14b, 14d, and 14e can be used to informally assess students' ability to reason about global information such as regularity, maximum/minimum values, slope, and variance as shown in line graphs.
- Problems 14c and 14d can be used to informally assess students' ability to extract local, point-specific information from line graphs.
- Problem 19 can be used to informally assess students' ability to recognize some of the limitations of line graphs.
- Problems 20 and 22 can be used to informally assess students' ability to tell the story depicted by a line graph in terms of the context.

B. FILLING IN TEMPERATURES

Mark's Project

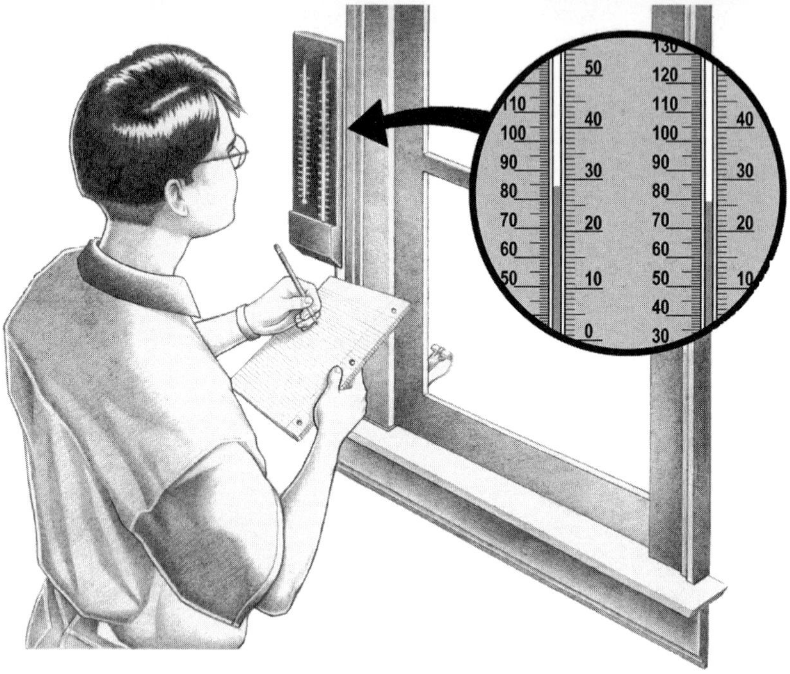

Mark has an indoor-outdoor thermometer at home that hangs near his window. The thermometer has a sensor that hangs outside so Mark can measure the temperature inside his room, as well as outside the house. As part of a school project, Mark reads the temperature several times during one weekend day.

	Outside	Inside
8:00 A.M.	60°F	65°F
10:00 A.M.	71°F	71°F
11:00 A.M.	76°F	75°F
1:00 P.M.	84°F	78°F
2:00 P.M.	90°F	78°F
3:30 P.M.	90°F	78°F
4:00 P.M.	88°F	78°F
6:00 P.M.	81°F	78°F

Mark put the temperatures in a table.

1. a. Was it ever warmer inside than outside?

   b. Why was the outside temperature not always the same?

   c. How is it possible that the temperature in Mark's room did not change much?

A Digital Thermometer

A Liquid Thermometer

Some thermometers show temperatures in digits. Other thermometers consist of a glass tube that contains mercury. As the temperature changes, the level of the mercury in the tube changes. Along the sides of the tube are number lines that allow you to read the temperature.

2. Read the temperature on the liquid thermometer on the left.

Solutions and Samples
of student work

1. **a.** yes, at 8:00 A.M.

   b. Answers will vary. Sample response:

   The sun started to shine and it was getting warmer during the day. At the end of the day, the sun went down and it cooled off.

   c. Answers will vary. Sample response:

   The house may have air-conditioning and his window was probably closed.

2. 84°F, 28.5°C

Hints and Comments

Overview Students compare inside and outside temperatures measured at different times. They read the temperature on a thermometer.

Planning You may use problem **1** as an introduction to the context of thermometers and temperatures. Students may work on problems **1** and **2** in pairs or small groups. Discuss students' solutions for both problems.

Comments about the Problems

1. Discuss the temperature range for one day in your area, both during the current time of year and in other seasons. This information can be used for this question as well as later when the students make their own graphs.
2. Encourage students to read the temperature in degrees Celsius as well as in degrees Fahrenheit.

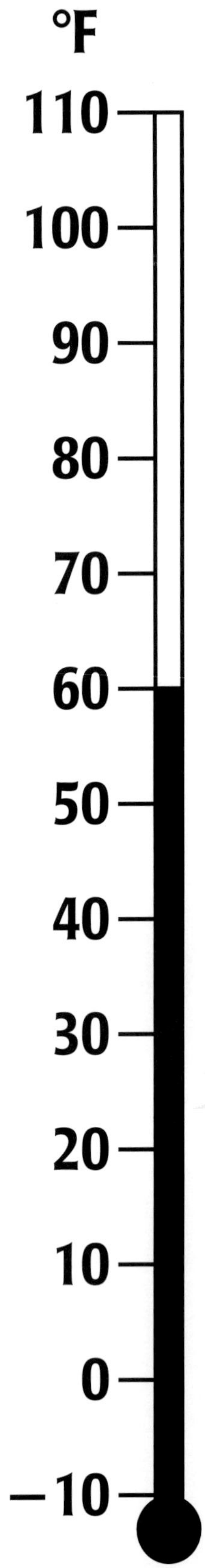

On the left is a thermometer showing the outside temperature at 8:00 A.M. This temperature was taken from Mark's table.

3. Using data from Mark's table, point to the levels of the liquid on the thermometer for the outside temperatures at 10:00 A.M. and 11:00 A.M.

	Outside	**Inside**
8:00 A.M.	60°F	65°F
10:00 A.M.	71°F	71°F
11:00 A.M.	76°F	75°F
1:00 P.M.	84°F	78°F
2:00 P.M.	90°F	78°F
3:30 P.M.	90°F	78°F
4:00 P.M.	88°F	78°F
6:00 P.M.	81°F	78°F

Cut out the thermometer strip on **Student Activity Sheet 5.** You can use it to make a graph of the outside temperatures during the day on the grid on **Student Activity Sheet 6.**

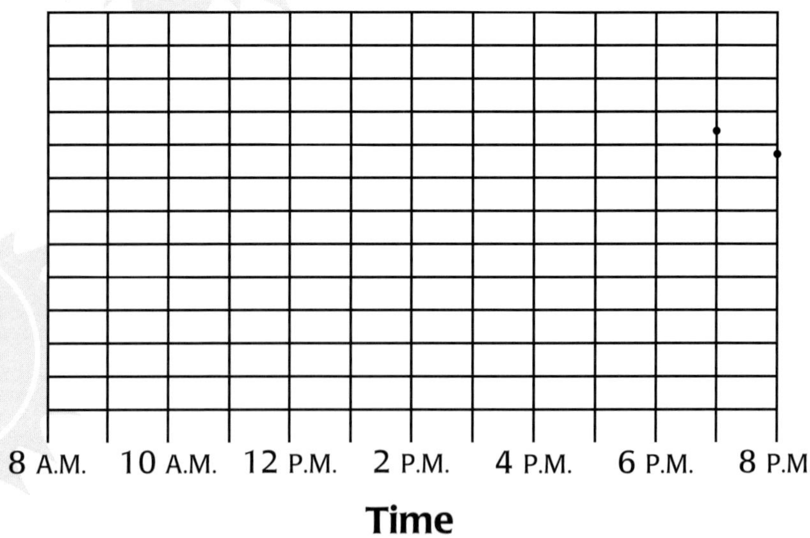

4. a. Put the thermometer strip on the grid at 8:00 A.M. Make a dot on the grid showing what Mark recorded as the outside temperature at that time. Do the same for all of the other outside temperatures in Mark's table.

b. Describe what you did to mark the outside temperature at 3:30 P.M. on the graph.

5. On the right side of the graph on **Student Activity Sheet 6,** there are already two dots. These are temperature measurements recorded later on the same day. They are not in the table. What are the temperatures, and at what times were they recorded?

3. 71°F and 76°F

4. a.

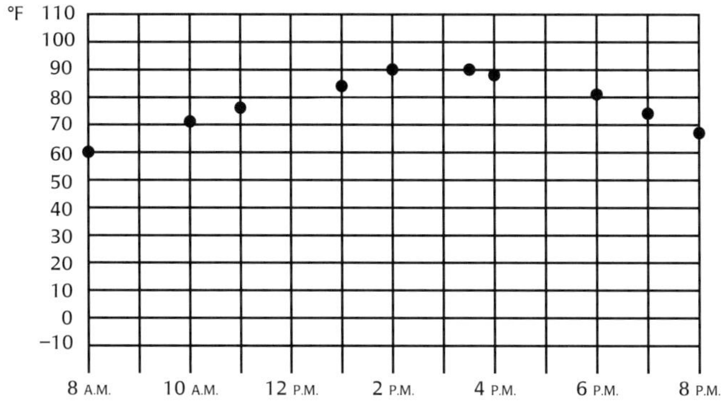

b. Descriptions will vary. Sample response:

The temperature, 90°F, is labeled on the thermometer strip, so I just followed that line over to halfway between 3 P.M. and 4 P.M. and put a dot there.

5. 73° or 74°F at 7:00 P.M.

67° or 68°F at 8:00 P.M.

Materials Student Activity Sheets 5 and 6 (one of each per student); scissors (one pair per student)

Overview Students use the data in the table to graph temperatures at different times during one day. They use a paper thermometer as a vertical axis to locate the correct points. They read the data for two points already drawn on the graph.

About the Mathematics On this page, students start to create a graph that shows the change in temperature during the day. They start in a way similar to the way they started in Section A. First they draw a couple of "snapshots" of the situation by putting dots on a graph. The proper height of a dot, which is related to a certain temperature, can be found by using a paper thermometer. Then, in problems **6, 7, 9,** and **10,** students add new dots to the diagram. They determine the locations of these dots by thinking about what is likely to occur between known data points. Or, in other words, students are interpolating data. By adding more and more dots, they eventually make a line graph.

Planning Students can work in pairs or in small groups on problems **3–5.** You may want to read ahead to problems **4, 6, 7, 9,** and **10** to see how the graph is going to develop. An effective way to show this change is to make an overhead transparency for each of the problems mentioned above. These transparencies can be laid on top of one another on the overhead projector to show how the graph evolves.

Comments about the Problems

4. a. This is a critical problem. Students should grasp the essential principles of constructing line graphs. Those who have seen graphs before might initially put scale numbers to the left of the graph and not use the paper thermometer. This is fine, *but do not ask this of the other students.* By moving from the concrete to the abstract, their understanding of graphs may be deepened.

5. Using the paper thermometer from Student Activity Sheet 5 can be of help here.

Mark did not measure the temperature at noon. From 11:00 A.M. to 1:00 P.M., the outside temperature rose by 8 degrees, from 76°F to 84°F. Mark assumed that it rose regularly by 2 degrees every half hour.

6. a. Was this a reasonable assumption? Why or why not?

b. Was it certain to have happened? Why or why not?

c. Using Mark's assumption, put dots on your graph on **Student Activity Sheet 6** for the temperatures at 11:30 A.M., 12:00 noon, and 12:30 P.M.

d. Could a warming trend of 2 degrees per half hour have continued after 1:00 P.M.? Why or why not?

Suppose Mark had measured the outside temperature every half hour of the day from 8:00 A.M. to 8:00 P.M. This would have resulted in more dots on the graph.

7. Put dots on your graph on **Student Activity Sheet 6** for every half hour that does not already have a dot. Be sure that these are likely temperatures. Use the paper thermometer if it is helpful.

Time	Temperature
12:00 NOON	
12:30 P.M.	
1:00 P.M.	84°F
1:30 P.M.	
2:00 P.M.	90°F
2:30 P.M.	
3:00 P.M.	
3:30 P.M.	90°F
4:00 P.M.	88°F

8. Copy the chart on the left. Fill in the missing temperatures from the graph you drew for problem **7.**

6. **a.** Yes, it is reasonable. Explanations will vary. Some students may say it is reasonable because a 2-degree rise every half hour results in an 8-degree rise in two hours.

b. No. Explanations will vary. There are several other possible scenarios. For example, the increase may have been greater in the first hour than in the second hour.

c.

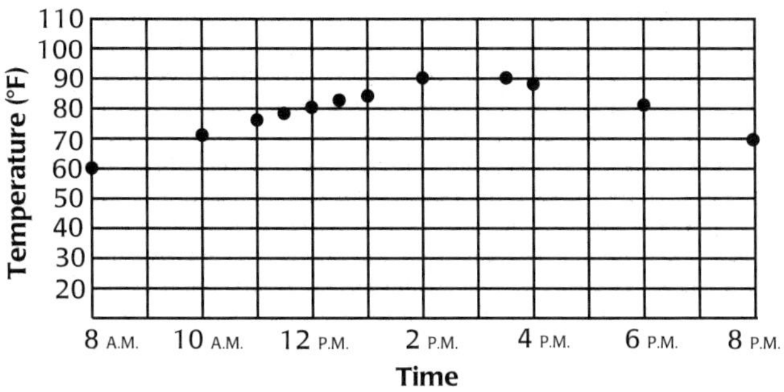

d. No. Explanations will vary. Some students may say that at 2:00 P.M. it was 90°F, not 88°F. However, the trend could have continued until 1:30 P.M., when it could have been 86°F.

7.

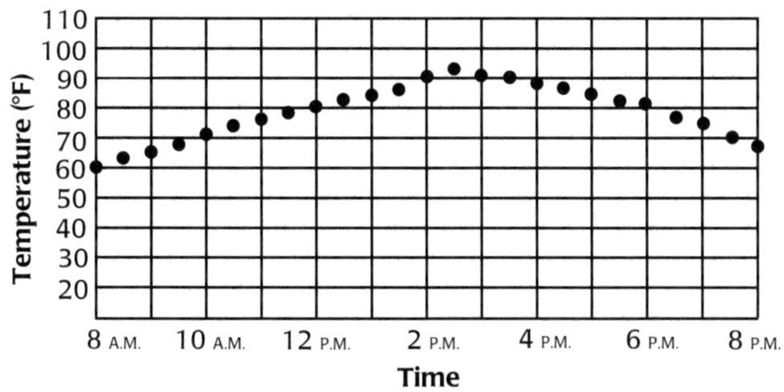

8. Answers will vary, depending on the positions of the points. Sample chart:

Time	Temp.
12:00 NOON	80°F
12:30 P.M.	82°F
1:00 P.M.	84°F
1:30 P.M.	87°F
2:00 P.M.	90°F
2:30 P.M.	92°F
3:00 P.M.	91°F
3:30 P.M.	90°F
4:00 P.M.	88°F

Materials Student Activity Sheet 6 (one per student); transparency of completed time/temperature chart, optional (one per class); overhead projector, optional (one per class)

Overview Students consider temperature changes and add more dots on their graph so that there is a point for every half hour increment of time. Then they read the data represented by the new points and put them in a table.

About the Mathematics Interpolating can be done in different ways. One way is to look at the data in the table, determine what might be reasonable data in between, and then draw points in the graph that represent these data. Another way is to look at the graph and use its shape to add new points.

Planning Students may work in pairs or in small groups on problems **6–8.** Discuss their answers briefly with the whole class. Focus the discussion on the assumptions that are reasonable for temperature changes. You may want to use a transparency of the completed time/temperature chart to show what their graph may look like.

Comments about the Problems

7. Students' answers should be similar to those shown in the Solutions column although the positions of the points may vary.

8. This exercise requires students to obtain information by reading and interpreting a graph. Answers are only correct if they fit the points in students' own graphs.

Mark could have measured the temperature every quarter of an hour.

9. Put dots on your graph for every quarter of an hour that does not already have a dot. Again, these should be likely temperatures.

10. a. On your graph, show what would happen if you put dots on the graph for measurements every minute.

b. Describe what your graph looks like now.

The smooth line that you get from adding more and more measurements is an example of a line graph.

11. Answer the following questions based on the line graph you just made:

a. How warm was it at 10:40 A.M.?

b. Name a time when it was 80°F.

c. What was the warmest temperature, and when did it occur?

You may have noticed that it is difficult to read information from the graph if you do not use the paper thermometer.

12. What could you add to your graph to read the temperatures without using the paper thermometer?

13. What differences do you see between this temperature line graph and the line graph of the race on page 10?

14. a. From the table on page 12, make a line graph for the temperature inside Mark's house. Use **Student Activity Sheet 6** again, but use a different color.

b. Describe some differences between the two temperature lines.

c. At what time was the temperature inside the house the same as outside?

d. When was it cooler inside than outside?

e. Was there more variation in the temperature inside or outside? How does this show up on the graph?

Solutions and Samples
of student work

9.

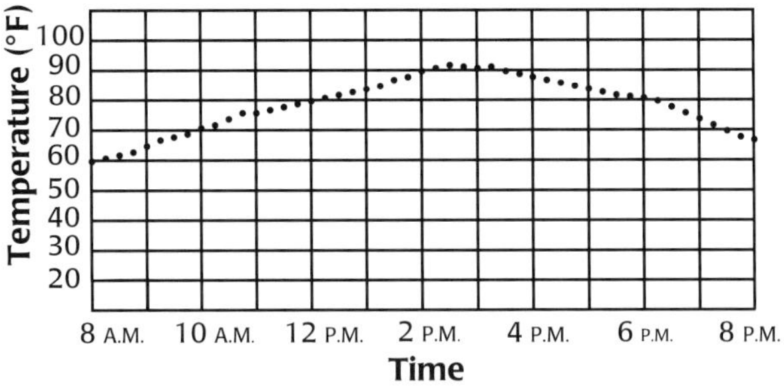

10. a–b. The dots are so close together that they appear to form a smooth curve.

11. a. 74°F or 75°F

b. around noon or a little after 6:00 P.M.

c. Answers will vary. Some students may say that the high temperature was about 90° or 92°F, and it occurred around 2:30 P.M.

12. You could write the temperatures to the left or to the right of the diagram.

13. Answers will vary. Sample responses:

- This line goes down after a while, which is not the case in any of the graphs of the race. In the race it would mean a runner would have to turn around and go backwards (not likely).
- Here you see the temperature changes over time (in hours). The race graphs show the distance over time (in minutes).

14. a. Line graphs will vary. Sample graph: between 8:00 A.M. and 8:00 P.M.

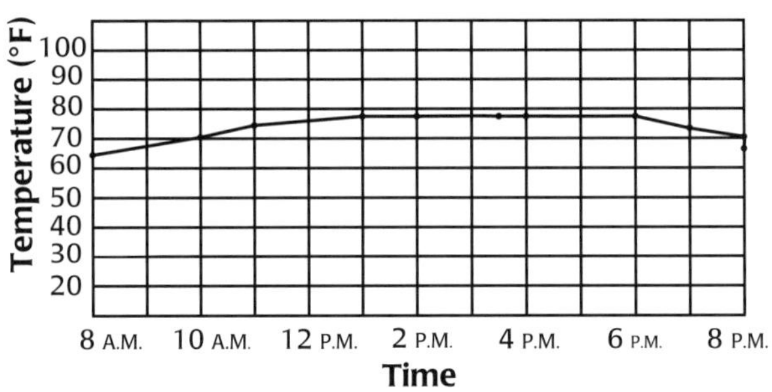

b. Descriptions will vary. Sample response:

The graph for the inside temperature is much flatter and does not go as high as the other graph.

c. 10:00 A.M.

d. any time after 10:00 A.M.

e. The outside temperature had greater variation. The flatter graph of inside temperatures indicates lesser variation.

Hints and Comments

Materials transparency of Student Activity Sheet 6, optional (one per class)

Overview Students add more points until they finally get a line graph of the outside temperatures. They compare this line graph and the line graph of the race. Then they create a line graph for the inside temperatures.

Planning Problems **10** and **11** are critical. Be sure to discuss these problems with the whole class. For a class discussion of problem **14,** you may want to use more transparency copies of Student Activity Sheet 6 to demonstrate how the graph can be made. Students may work in pairs or small groups on problems **9–13.** You may want students to work individually on problem **14.** Problem **14** can be used as informal assessment.

Comments about the Problems

9. The graph shown in the solutions column is intended to be a sample graph. Students' graphs will vary, depending on their results from problem **7.**

Note: The temperature peaks between 2:00 P.M. and 3:30 P.M. This time period is investigated in detail on Student Book page 17.

10–11. These problems are critical because in them students will grapple for the first time with a line graph's property of being continuous.

11. Reading with the help of the paper thermometer is still appropriate.

14. Informal Assessment Problem **14a** assesses students' ability to create line graphs from numeric data, and to interpolate data points to fill in line graphs when appropriate. Problems **14b, 14d,** and **14e** assess students' ability to reason about global information such as regularity, maximum/minimum values, slope, and variance as shown in line graphs. Problems **14c** and **14d** assess students' ability to extract local, point-specific information from line graphs.

15. Imagine that the opening in the graph mask is a thermometer. What would you see if you pulled the temperature line graph through the mask?

The day after Mark made his graph, he read in the morning paper that 70° was the average temperature for the previous day. Mark was surprised, because he thought that the average temperature was warmer than 70°.

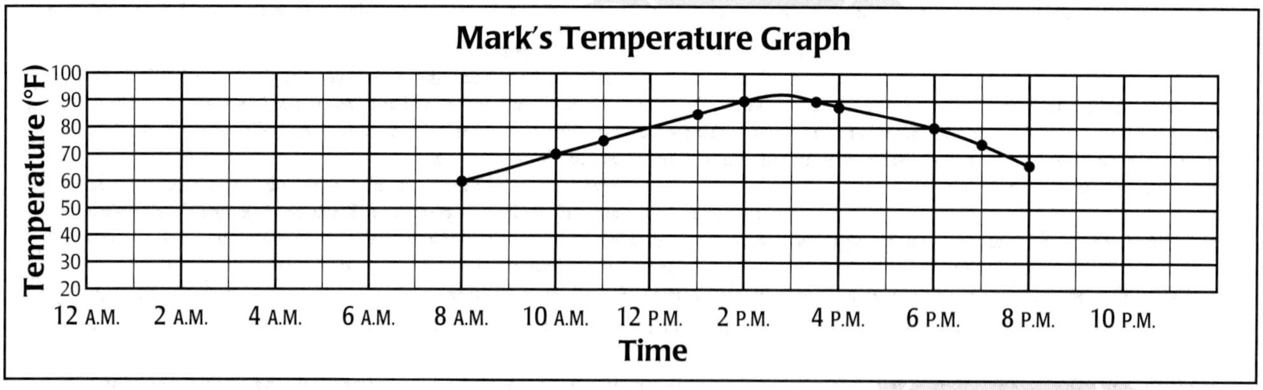

The above graph shows the outside temperatures Mark collected.

16. a. On the copy of the graph on **Student Activity Sheet 7,** draw a horizontal line to represent the average temperature as reported by the newspaper.

b. Why do you think that Mark thought the average temperature should be higher? Draw a second horizontal line on **Student Activity Sheet 7** to show the average temperature for Mark's data.

c. Explain how the newspaper could have been correct.

15. You would see the top level of liquid in the thermometer moving up and down.

16. a.

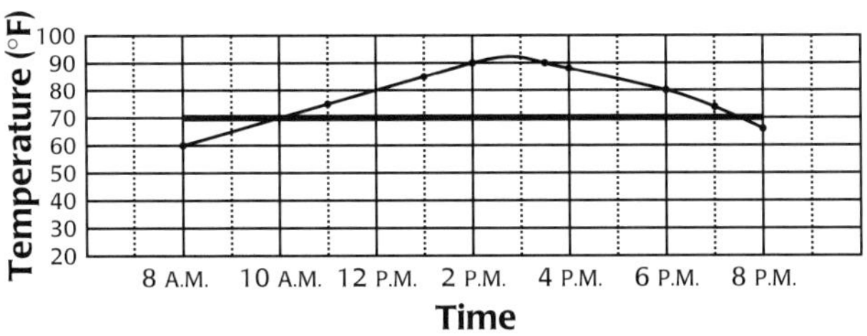

b. Explanations will vary. Some students may say that Mark thought the average temperature was higher because almost all of his graph is above the 70° mark.

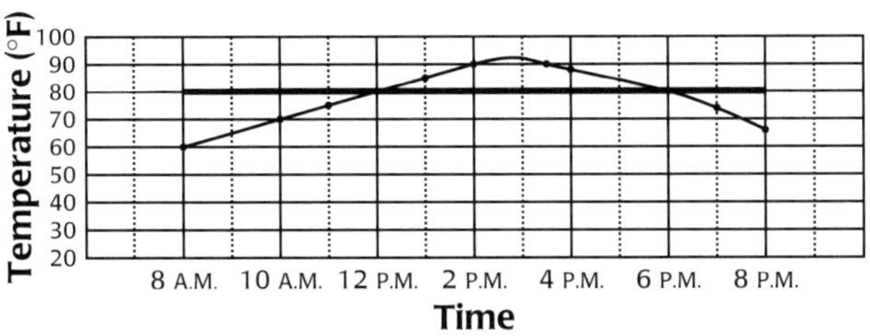

c. Explanations will vary. Some students may reason that the newspaper probably included temperatures for the morning, from midnight to 8:00 A.M., and again for the evening, from 8:00 P.M. to midnight. Those temperatures were probably cooler than during the day, and they would lower the average temperature.

Hints and Comments

Materials Student Activity Sheet 7 (one per student)

Overview Students use a graph to investigate the average temperature.

About the Mathematics In the grade 5/6 unit *Picturing Numbers,* students studied the concept of average. Two different strategies are taught in that unit. The most common strategy used to compute the mean is the conventional algorithm: dividing the sum of a set of terms by the number of terms.

The other strategy is the *compensation strategy,* in which one finds the mean by redistributing the numbers. For temperatures, this would mean that the higher temperatures would be made lower, and the lower temperatures should then rise by the same number of degrees. The mean can be "estimated" by using a graph in which one draws a horizontal line so that the parts above the line may be used to fill the gaps below the line.

Planning Students may work on problems **15** and **16** individually. When students have finished problem **15,** you may wish to discuss their answers and briefly review the concept of average before students continue.

Comments about the Problems

16. a. Notice that there is a temperature scale on the given graph, so students do not need a paper thermometer.

HIGH FOR THE DAY

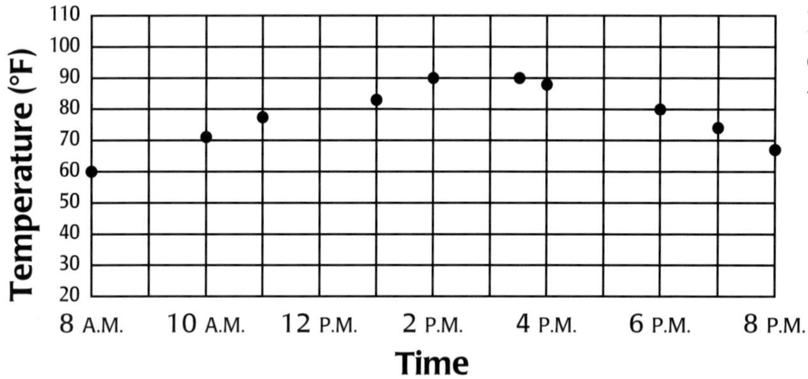

Look at the dots in the diagram on the left showing the temperatures from Mark's table.

Below are three possibilities for what might have happened to the temperature between 2:00 P.M. and 3:30 P.M.

17. For each possibility, fill in the line graph between 2:00 P.M. and 3:30 P.M. and record the high. Use the three graphs on **Student Activity Sheet 8.**

Possibility a
Between 2:00 P.M. and 3:30 P.M., the temperature stayed at 90°F.

Possibility b
Between 2:00 P.M. and 3:30 P.M., it got even hotter than 90°F.

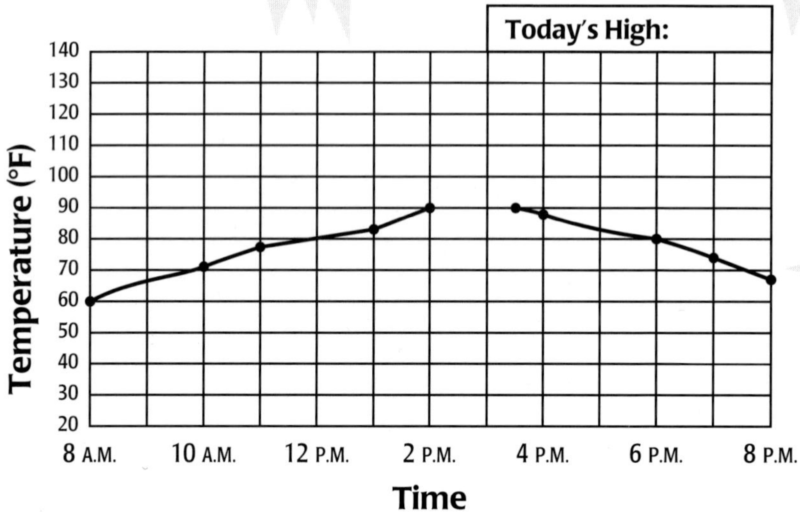

Possibility c
After 2:00 P.M., there was a severe thunderstorm, which caused the temperature to drop. When the storm was over, the temperature rose again. At 3:30 P.M., it was again 90°F.

Solutions and Samples
of student work

17.

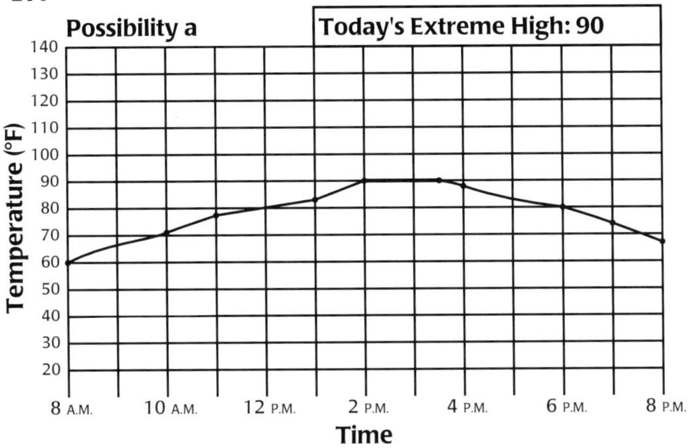

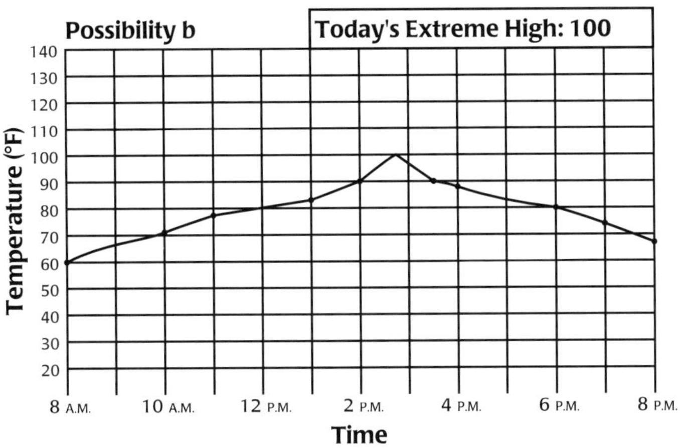

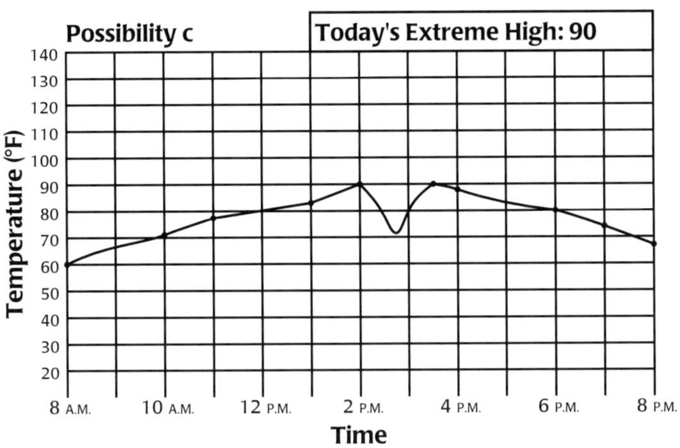

Hints and Comments

Materials Student Activity Sheet 8 (one per student); transparency of Student Activity Sheet 8, optional (one per class); overhead projector, optional (one per class)

Overview Students complete the missing part of a graph in three different ways, according to given scenarios.

Planning Students can work in pairs or in small groups on problem **17.** You can make a transparency of Student Activity Sheet 8 to illustrate the possible solutions of problem **17.**

Comments about the Problems

17. The purpose of this problem is for students to focus on interpolation and to reflect on what is possible and likely and what is not. For possibilities **b** and **c,** students might exaggerate by drawing huge peaks. Make sure they know that such situations are unlikely. The temperature might even fluctuate up and down although that is not likely in a period of one and a half hours.

You have looked at three possibilities for the outside temperatures between 2:00 P.M. and 3:30 P.M. and the corresponding graphs.

18. Are there any other possibilities? If so, how many?

Each graph below shows just the part of the graph between noon and 5:00 P.M. Every graph shows a different situation between 2:00 P.M. and 3:30 P.M.

19. Explain why these situations are not likely to have occurred.

a.

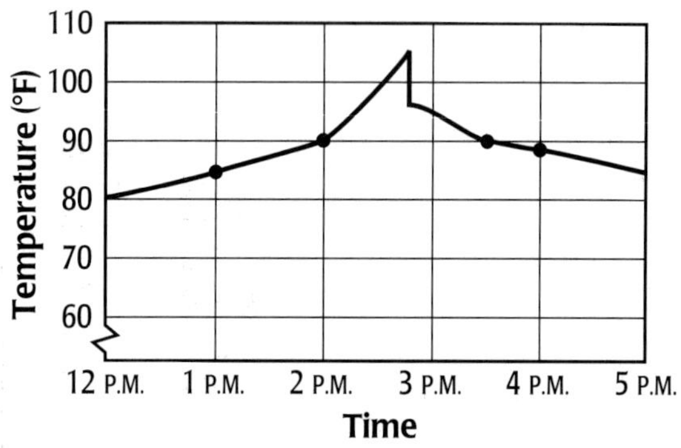

b.

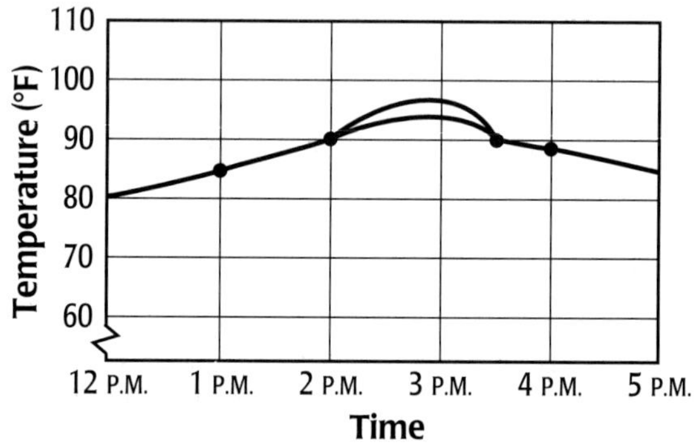

c.

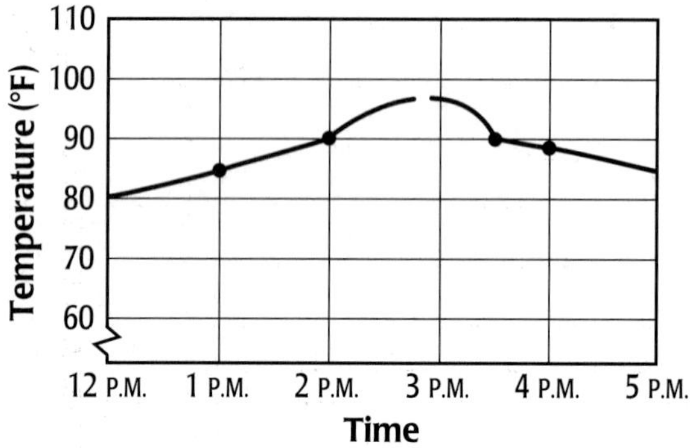

d.

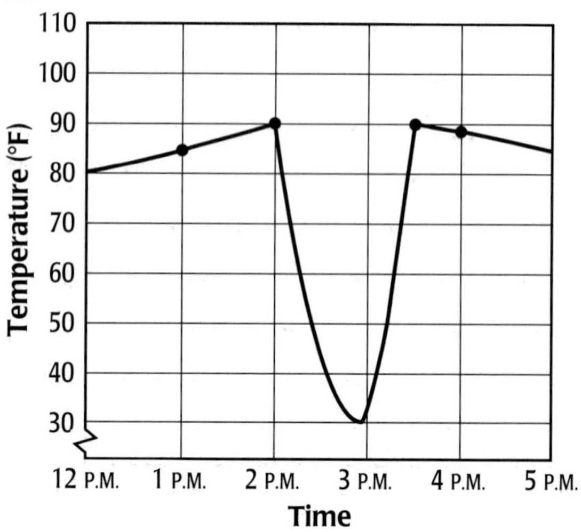

Solutions and Samples
of student work

18. Answers will vary. Sample student response:

No. The only possibilities are higher temperatures, lower temperatures, and the same temperature. What can vary are the actual high and low.

19. a–d. Explanations will vary. Sample explanations:

a. The drop at 2:40 P.M. is too sudden. It appears to drop 10° in one minute.

b. There cannot be two temperatures for the same time.

c. There seems to be no temperature for 10 minutes. There should be dots for each moment in between 2:00 P.M. and 3:00 P.M.

d. The change in temperature is very unlikely, since the drop is too deep. Outside temperature cannot drop from 90°F to 30°F and rise from 30°F to 90°F in $1\frac{1}{2}$ hours.

Hints and Comments

Overview Students investigate four examples of completing the graph that show the possible temperature fluctuations between 2:00 P.M. and 3:30 P.M. They explain for each situation why each temperature change is not likely to have occurred.

About the Mathematics Some of the situations shown in problem **19** are impossible. Graphs of situations that deal with a continuous process of change over time do not have "holes." And it is not possible for there to be two values at a single time.

The study of graphs showing change over time is continued in the grade 7/8 unit *Ups and Downs* and in the grade 8/9 unit *Growth.* In these units, important features of graphs like maximum, minimum, and slope are further investigated. The concept of slope is revisited in the grade 7/8 unit *Looking at an Angle.*

Planning You may want to start with a class discussion about the meaning of the jagged line on the vertical axis, which indicates that the axis does not begin at zero. This would also be a nice opportunity to review the concepts of horizontal and vertical axes. Students can work in pairs or in small groups on problems **18** and **19.** Have students share their explanations in their group or with the class. Problem **19** can be used as informal assessment. The Extension may be assigned as homework.

Comments about the Problems

19. Informal Assessment This problem assesses students' ability to recognize some of the limitations of line graphs.

Extension You might ask your students to make up three situations in the context of the race that are not likely to have occurred and to graph them. This task can also be used as informal assessment and assigned as homework. It assesses the same goal as problem **19.**

The *Weather* Journal

During one hot week of summer vacation, Amber made a weather journal.

Monday: It was a beautiful day. The sun was shining all morning and it got fairly hot. We decided to go hiking but did not take our jackets. Suddenly, the wind started to blow and it got chilly. It even rained for a while. Lucky for us, the sun came back, and our clothes could dry.

Tuesday: It was sunny all day today. It seemed warmer than yesterday. According to the weather forecast, tomorrow will be even hotter.

Wednesday: Today was hot, even blistering at times. At 5:00 P.M., a thunderstorm finally cooled things off.

Thursday: Yesterday's thunderstorm made a mess of the good weather. It rained all day, with temperatures below 65°.

Friday: Today was partly cloudy. There were waves of warm and cold temperatures. The last warm period was the longest.

Saturday: The weather was good all day, except around 11:00 A.M. During the afternoon, it was partly cloudy, but it never got really cold.

Solutions and Samples
of student work

Hints and Comments

Materials weather data illustrated in graphs from local newspapers, optional (two or three graphs per class)

Overview Students read weather journals compiled over a period of five days. There are no problems on this page for students to solve.

Planning You may want to have students work on the Writing Opportunity suggested below as homework and then share their work with other students.

Writing Opportunity Bring in local weather charts, such as from a newspaper, and have each student write a story based on the graphs. This activity may be assigned as homework.

Amber also made a temperature graph for each day.

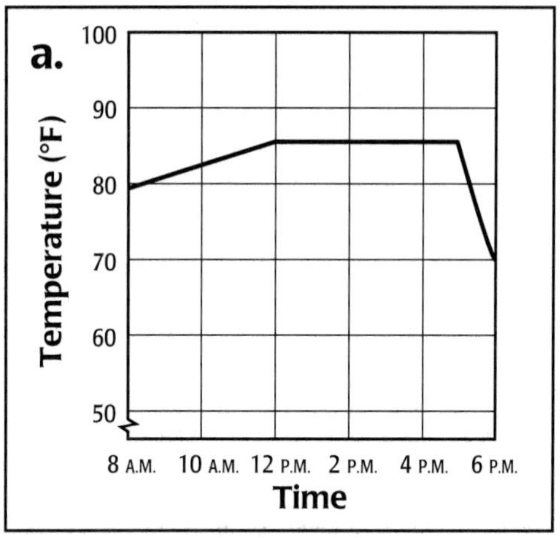

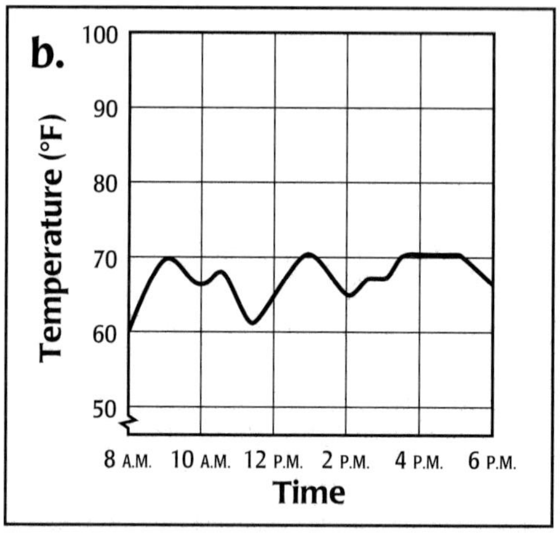

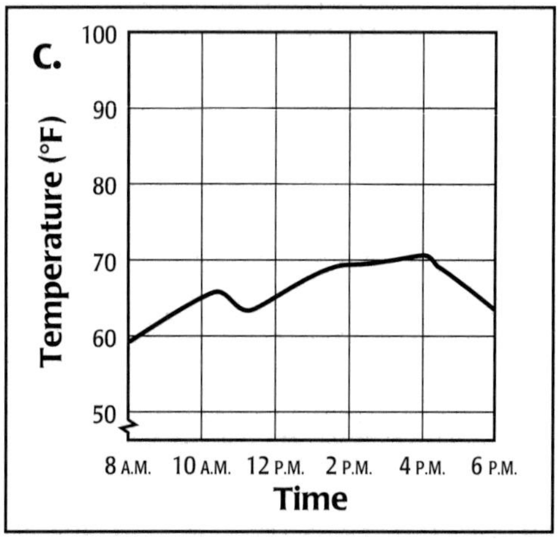

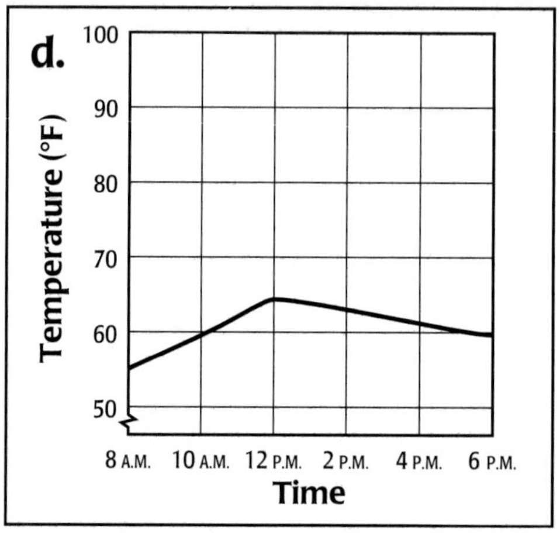

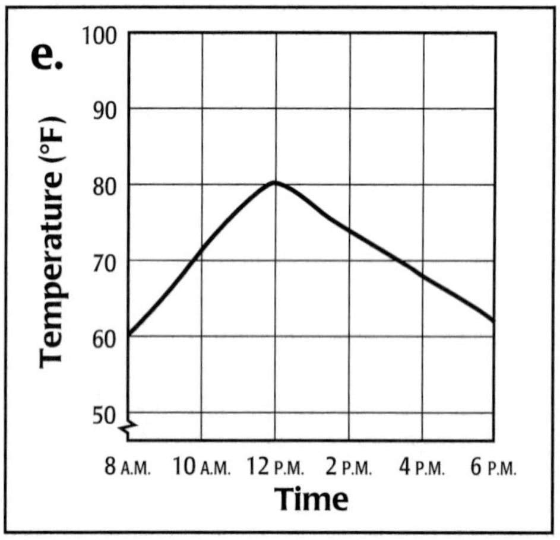

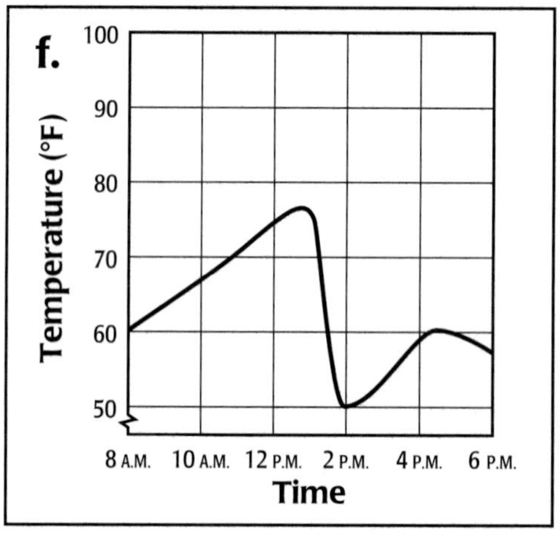

20. Which graph belongs to which day? How do you know?

20. Graph **a:** Wednesday
Graph **b:** Friday
Graph **c:** Saturday
Graph **d:** Thursday
Graph **e:** Tuesday
Graph **f:** Monday

Explanations will vary. Sample explanations:

Monday has to be graph **f** because graph **f** shows a sudden temperature drop before it then gets warmer again.

Tuesday is graph **e,** a nice warm day.

Wednesday is graph **a** because it shows a hot day with a sudden temperature drop at 5:00 P.M. (Notice that graph **a** shows a hotter day than graph **e.**)

Thursday has to be graph **d**—a cool day with temperatures below 65°.

Friday has to be graph **b** because this graph shows lots of ups and downs with one long warm period.

Saturday is graph **c.** It shows a nice day with some cooling off before noon, and then it gets warmer again.

Overview Students match each of five different graphs with one of the stories on Student Book page 18.

Planning You may want to have students work individually or in pairs on problem **20.** This problem can be used as informal assessment.

Comments about the Problems

20. Informal Assessment This problem assesses students' ability to tell the story depicted by a line graph in terms of the context.

This problem may encourage students to look more globally at significant shapes in the graphs (decrease, increase, maximum, minimum, and constancy).

Ask students how they organized their work and how they matched the graphs and the stories. Did they read a story and then look for a matching graph, or vice versa? What pair did they find first?

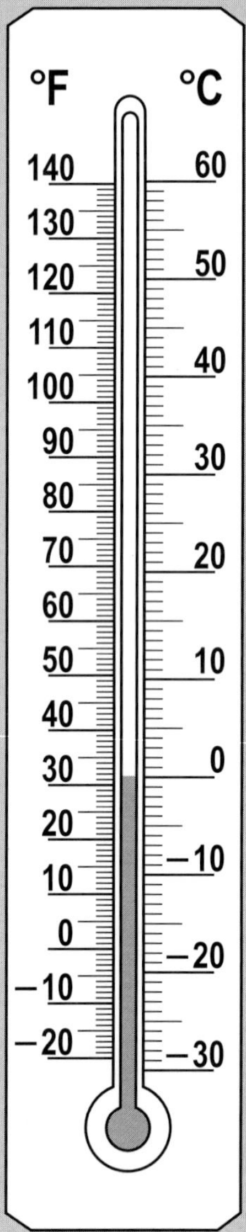

Each of the graphs on the previous page had a temperature scale from 50 to 100 degrees Fahrenheit. Temperature graphs can also show degrees Celsius.

The table on the right shows some temperatures in degrees Fahrenheit recorded every two hours for 10 hours.

Time	°F	°C
8 A.M.	60	
10 A.M.	67	
12 P.M.	75	
2 P.M.	50	
4 P.M.	58	
6 P.M.	57	

21. a. On **Student Activity Sheet 9,** fill in what each temperature would be in degrees Celsius. Use the thermometer on the left to help you.

b. Use the grid on **Student Activity Sheet 9** to make a graph of the Celsius temperatures.

c. The temperatures in the table above came from one of the six graphs on page 19. From which graph do you think it came? Explain how you know.

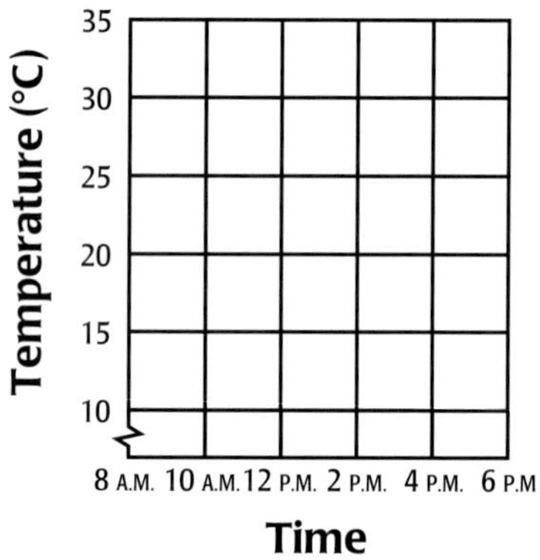

Summary

Line graphs can show how things change over time.

For example, in this section you made line graphs to show how the temperature changed during a day.

If you are given only a few points on a graph of temperature readings for one day, there are many possible line graphs.

Having more information, such as when it was stormy, helps you make temperature line graphs more accurate.

Summary Questions

Time	°F
9:00 A.M.	60°
12:00 P.M.	75°
3:00 P.M.	84°
6:00 P.M.	79°

Consider the table of temperatures on the left.

22. Draw a line graph using these data.

23. Draw another possible line graph based on the same data but with the additional information that there was a thunderstorm at 1:00 P.M.

21. a. Answers will vary. Sample table:

Time	°F	°C
8 A.M.	60	15
10 A.M.	67	18
12 A.M.	75	23.5
2 P.M.	50	11.5
4 P.M.	58	14
6 P.M.	57	13.5

b. Graphs will vary. Sample graph:

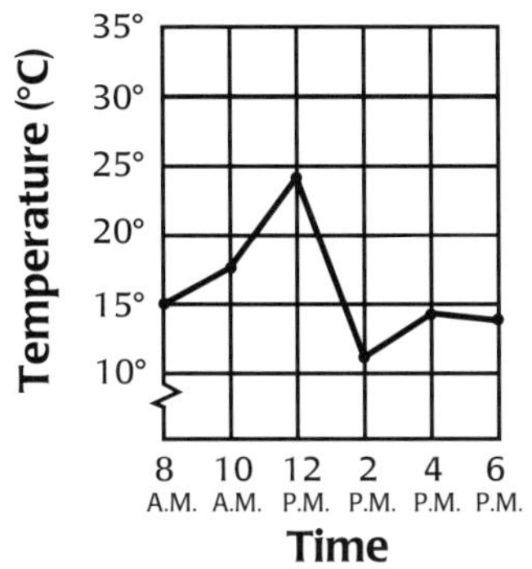

c. The data is from graph **f.** Explanations will vary. Students may check using either the original Fahrenheit data or their new Celsius graphs.

22.

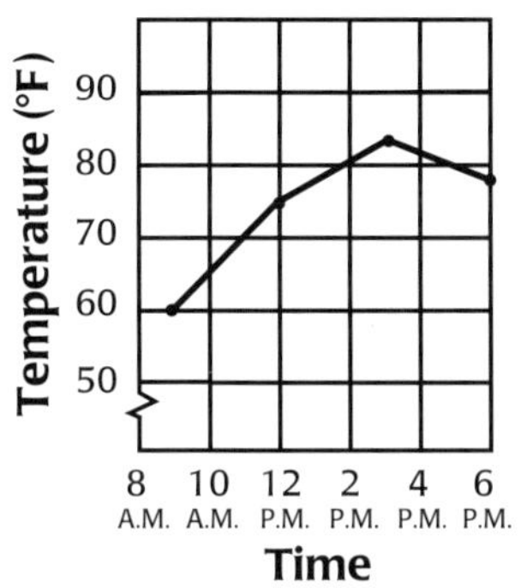

23. Graphs may vary. Sample graph:

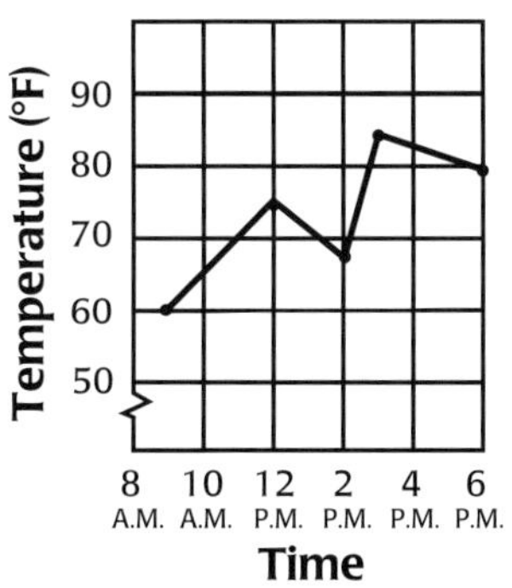

Materials Student Activity Sheet 9 (one per student); graph paper (one sheet per student)

Overview Students use the scales on the thermometer to convert temperatures from Fahrenheit to Celsius. Then they make a graph of the Celsius temperatures.

Students read the Summary and complete two problems involving making line graphs using numeric data.

About the Mathematics A liquid thermometer with two scale lines, one in degrees Celsius and the other in degrees Fahrenheit, shows how these temperatures are related; a rise of 10°C increases 18°F. These two scale lines do not show a double number line with equal ratios. This can be illustrated with the following example: When the temperature of 10°C doubles into 20°C, the temperature in °F will not double (10°C = 50°F; 20°C = 68°F). In the grade 7/8 unit *Building Formulas,* students will investigate the conversion formulas.

Planning Have students work in pairs or in small groups on problems **21–23.** Discuss these problems with the whole class. Problem **21** may be assigned as homework. Problem **22** can be used as informal assessment.

Comments about the Problems

21. Homework This problem may be assigned as homework.

c. This problem will focus on the shape of the graph rather than on the data represented. Note that the table does not show the maximum temperature at 1:00 P.M. You may want to have a class discussion about the advantages and disadvantages of using a table in comparison to a graph.

22–23. If students are having difficulty, you may encourage them to use previous graphs as examples. Different scales will affect the shape of the graph. For example, a scale that uses intervals of 10°F will make the graph look more "squat" than a scale that uses steps of 5°F.

22. Informal Assessment This problem assesses students' ability to tell the story depicted by a line graph in terms of the context.

Work Students Do

Students create week-long temperature graphs by squeezing the data from individual daily temperature graphs. Students continue this theme of compressing graph data to make month-long and year-long temperature graphs. At the end of the section, students investigate a year-long graph of temperatures in a city in France.

Goals

Students will:

- extract local, point-specific information from line graphs;
- reason about global information such as regularity, maximum/minimum values, slope, and variance as shown in line graphs;
- understand limitations of line graphs;
- recognize the importance of scale in creating line graphs.

Pacing

- approximately two 45-minute class periods

About the Mathematics

This section explores the loss of detail and the gain of a more global picture in graphs that cover longer and longer periods of time but remain the same size. For example, when the week-long graph is compressed to become part of a month-long graph, the time scale on the horizontal axis changes so that each square represents six hours rather than one hour. When students draw or interpret compressed graphs, they will focus more on the global features, such as minimum and maximum values, slope, and variance. Compressing the time scale leads to a loss of information detail, since individual points are more difficult to read from a compressed graph.

Materials

- Student Activity Sheets 10–12, pages 122–124 of the Teacher Guide (one of each per student)
- scissors, page 55 of the Teacher Guide (one pair per student)
- tape or glue, page 55 of the Teacher Guide (one roll or bottle per pair of students)
- transparency of year-long graph, page 59 of the Teacher Guide, optional (one per class)
- overhead projector, page 59 of the Teacher Guide, optional (one per class)

Planning Instruction

You may want to start this section with a class discussion about the seven graphs on Student Activity Sheets 10 and 11. You could also ask students questions that invite them to look globally at the graphs before they begin to study them more in detail. You might want to ask students, *What do all seven graphs have in common?* [They all show how temperature changes over a whole day; the temperature first goes down, then up, and then down again.] *Which graphs do you find the most remarkable? Why?* [Possible answers include Thursday because of the sudden drop; Sunday because it shows little difference between high and low temperatures.]

Students may work on problem 1 in small groups or in pairs. They can work on problems 2, 3a–c, 4, 5, and 10 in pairs and problems 6–9 individually or in pairs. Students can work on problem 3d individually. Problem 3d may also be assigned as homework.

There are no optional problems in this section.

Homework

Problems 3d (page 56 of the Teacher Guide) and 4 and 5 (page 58 of the Teacher Guide) may be assigned as homework. The Writing Opportunity (page 57 of the Teacher Guide) may also be assigned as homework. After students complete Section C, you may assign appropriate activities from the Try This! section, located on pages 39–42 of the *Tracking Graphs* Student Book. The Try This! activities reinforce the key mathematical concepts introduced in this section.

Planning Assessment

- Problem 1 can be used to informally assess students' ability to extract local, point-specific information from line graphs.
- Problems 3d and 10c can be used to informally assess students' ability to understand limitations of line graphs and to recognize the importance of scale in creating line graphs.
- Problem 6 can be used to informally assess students' ability to reason about global information such as regularity, maximum/minimum values, slope, and variance as shown in line graphs.

C. GRAPHS OVER A LONGER TIME PERIOD

One **Day** *at a Time*

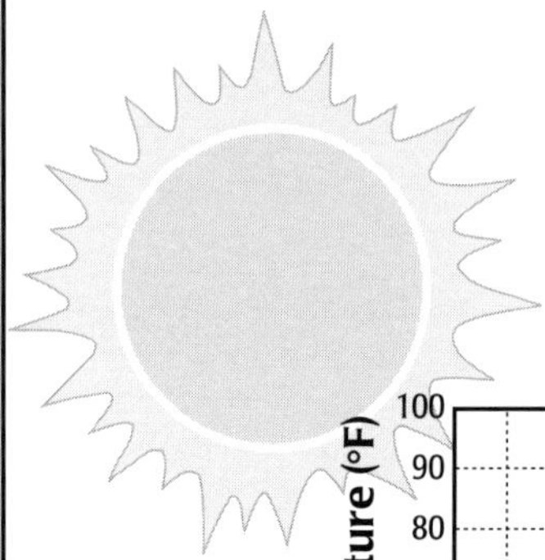

Student Activity Sheets 10 and **11** show temperature graphs for the first seven days of June in the mountains.

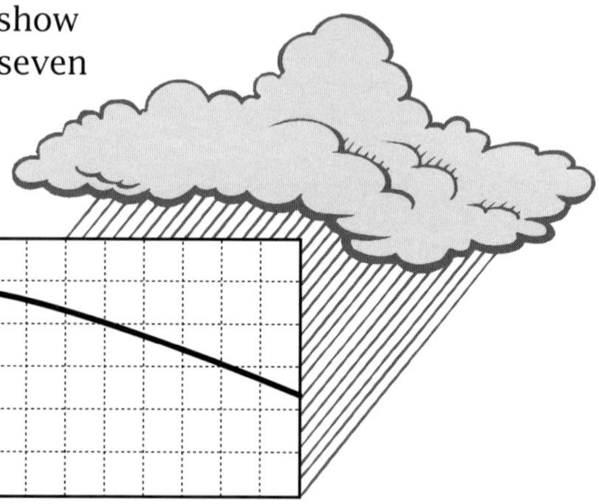

Monday, June 1st

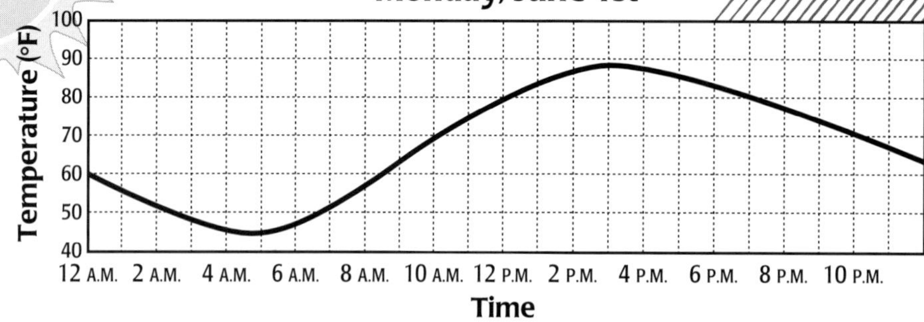

Tuesday, June 2nd

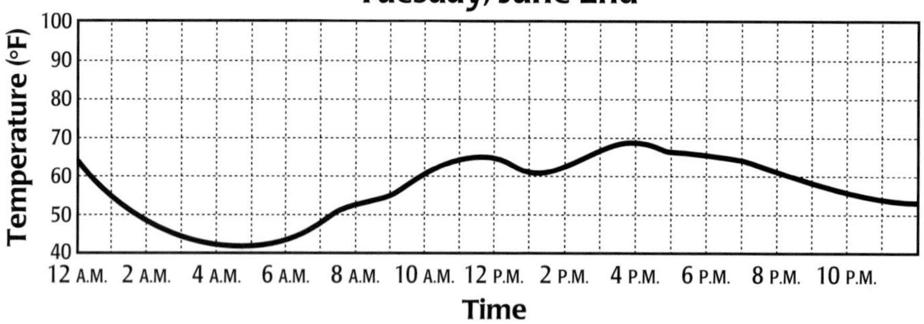

Wednesday, June 3rd

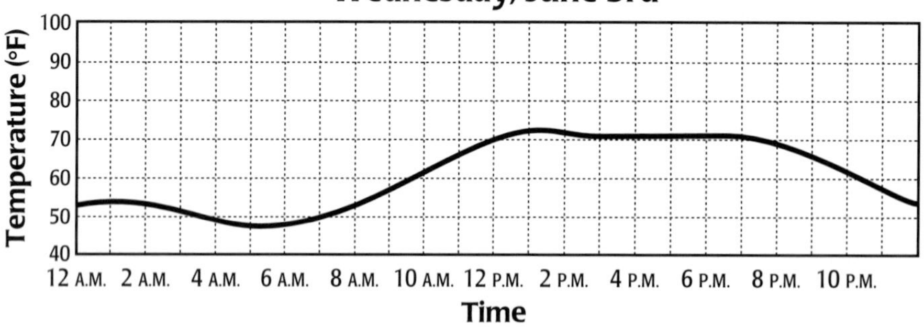

Thursday, June 4th

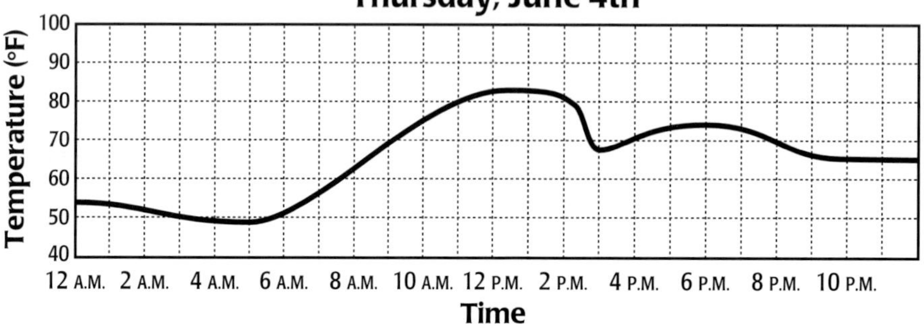

Solutions and Samples
of student work

Hints and Comments

Materials Student Activity Sheets 10 and 11 (one of each per student)

Overview Students are introduced to temperature graphs. There are no problems on this page for students to solve.

Planning You may wish to discuss with students whether the graphs on Student Activity Sheets 10 and 11 closely follow the temperature data for a week where you live, and if not, then what such graphs might look like.

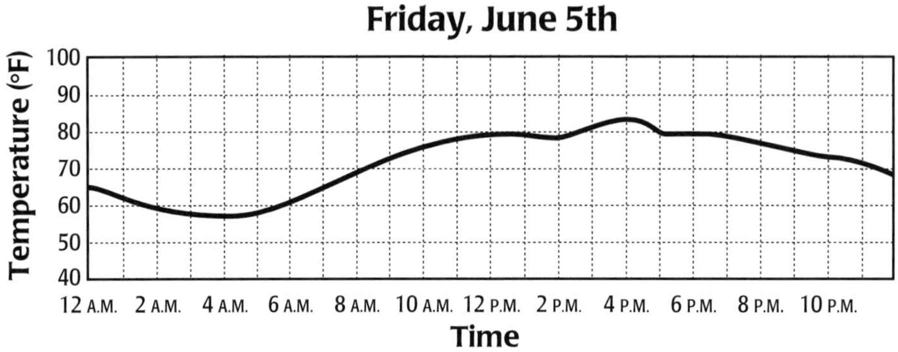

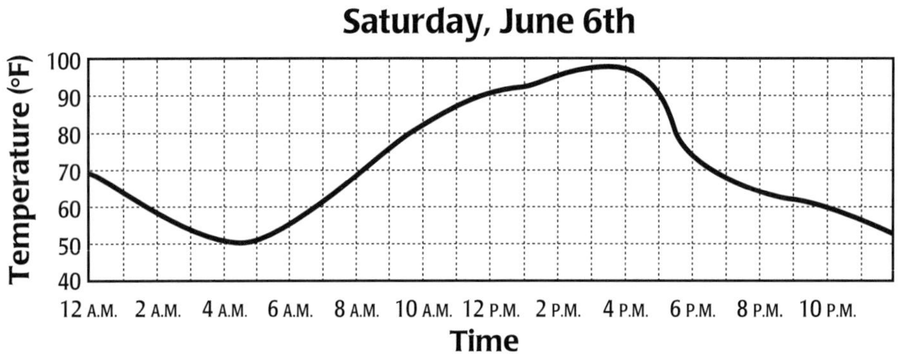

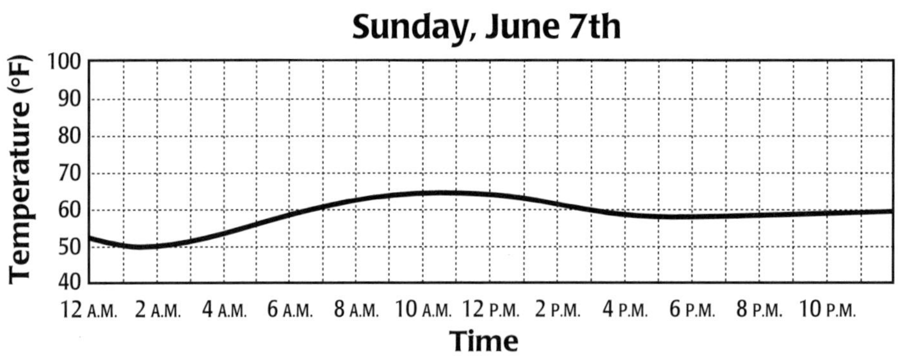

1. a. What was the temperature on Tuesday at 10:00 A.M.?

b. There are two places where you can find the temperature at the beginning of Friday. Where are they?

c. What was the hottest temperature during this week, and when did it occur?

d. What was the coldest temperature during this week, and when did it occur?

Solutions and Samples

of student work

1. **a.** 60°F

   b. to the far right of the Thursday graph (at 12 midnight) and to the far left of the Friday graph (at 12 midnight)

   c. about 97°F, Saturday, June 6, around 3:45 P.M.

   d. about 42°F, Tuesday, June 2, around 4:30 P.M.

Hints and Comments

Overview Students investigate seven day-long temperature graphs. They locate specific points and compare the graphs to find the highest and lowest temperatures of the week.

Planning Students can work in pairs or in small groups on problem **1.** This problem may be used as an informal assessment. Discuss students' solutions for problem **1.**

Comments about the Problems

1. **Informal Assessment** This problem assesses students' ability to extract local, point-specific information from line graphs.

   b. The temperature at 12 A.M. Friday is 65°F. Discuss the fact that the day-long graphs link up (the next one starts where the previous one ends) and why. You might ask students whether this is also true for the temperatures at 12 A.M. on the other days. This problem may induce students to suggest that it is possible to put the graphs next to each other.

Looking *at a Week*

It would be nice to show the temperatures for the whole week in a single graph. To do this, you could start with a long, thin grid and draw the line graph on it, or you could tape together the graphs you already have.

2. Cut out the individual graphs on **Student Activity Sheets 10** and **11** and glue or tape them together. Each graph is about 12 centimeters long. About how long is your graph for the whole week?

The difficulty with this graph for the entire week is that it is too long! You cannot put it in your notebook unless you fold it up, and you cannot see the whole graph at a single glance.

To make the week-long graph fit on a normal-size page, you would have to squeeze it.

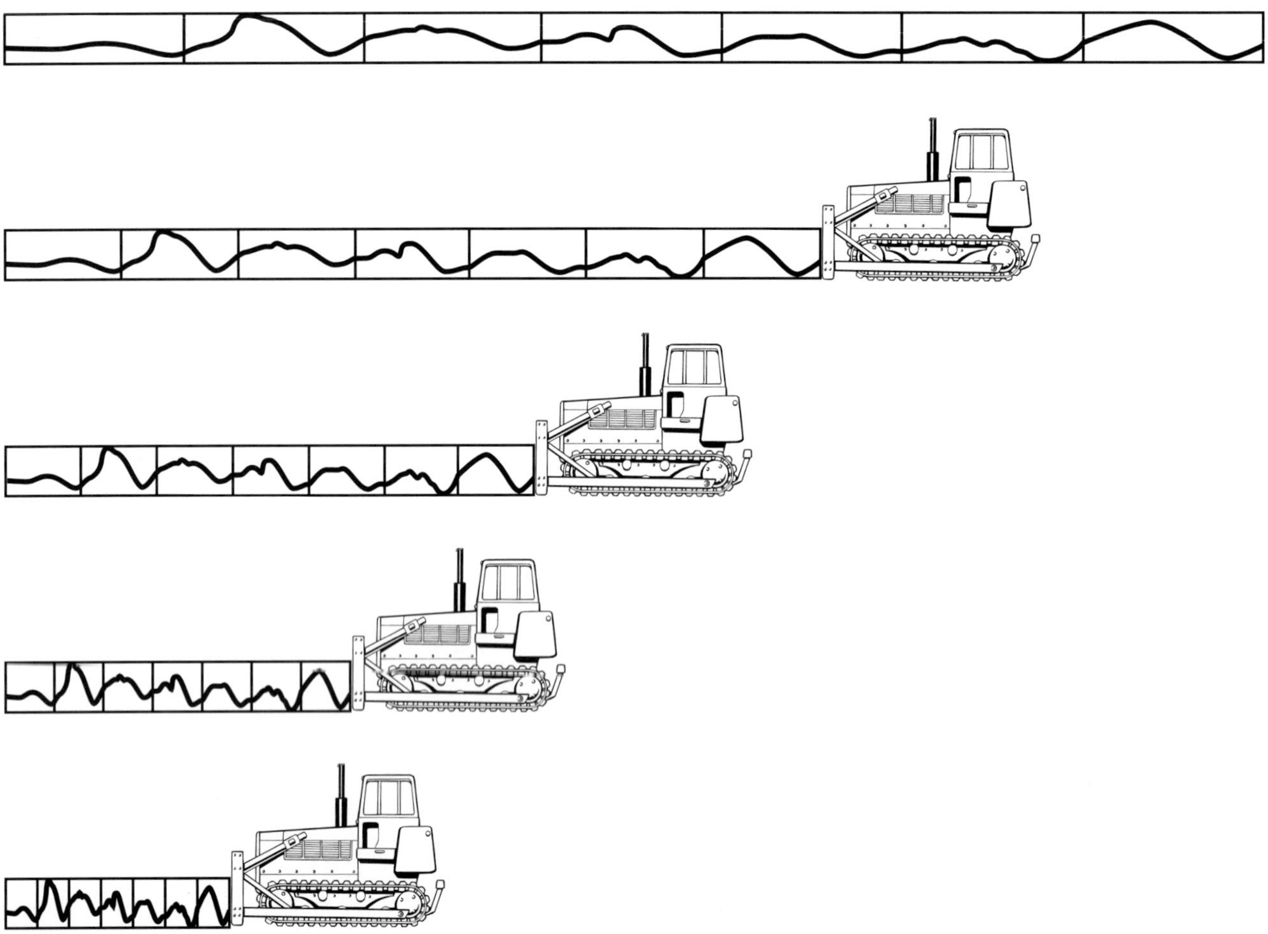

Solutions and Samples
of student work

2. The week-long graph is 84 centimeters (12 centimeters per day times seven days).

Hints and Comments

Materials scissors (one pair per student); tape or glue (one roll or bottle per pair of students)

Overview Students cut out the seven graphs and tape or glue them together to make a week-long graph.

About the Mathematics One of the purposes of the activities in this section is to focus students' attention to scale. "Squeezing" the graph is, mathematically speaking, a way of transforming the horizontal axis through multiplication so that the unit of the time scale changes into a larger unit of time. Before compressing, the time unit (one square) was 1 hour. After compressing (see Student Book page 24), the unit (one square) represents 6 hours. Students should understand the idea of squeezing the graph and its effect on the horizontal time scale.

Planning Students can work in pairs on problem **2.** It might save time if each pair of students cuts out and tapes or glues together one series of graphs.

You may want have a short class discussion about "squeezing" the graphs. Compression of a graph can also be shown with a broad rubber band: When a graph is drawn on a stretched-out rubber band, the compressed graph shows up when the rubber band is released.

Comments about the Problems

2. If students measure the length of the graph, they might have missed the information in the text about the length of each one-day graph.

3. **a.** Use the long graph that you made in problem **2** to help you draw a week-long graph on **Student Activity Sheet 12.** Be as accurate as you can.

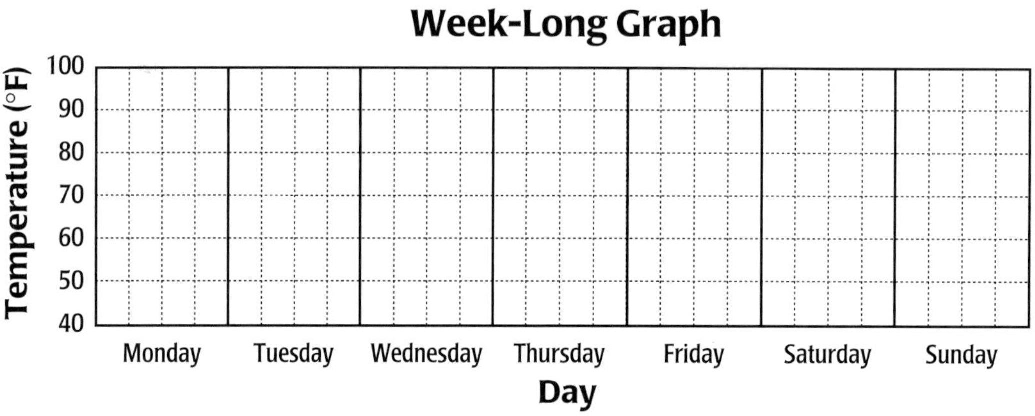

b. In this week-long graph, can you still find what the temperature was on Tuesday at 10:00 A.M.? Why or why not?

c. Can you see what Tuesday's highest temperature was? Why or why not?

d. Explain in writing the advantages and disadvantages of the week-long graph compared with the seven day-long graphs.

Instead of looking at the temperature over one week, you might want to look at temperatures for the whole month. Below is a description of the weather for the next three weeks of June.

The second week was a bit colder than the first week.

The third week was even colder.

The fourth week showed a return to warmer temperatures.

3. a. Graphs will vary. Sample graph:

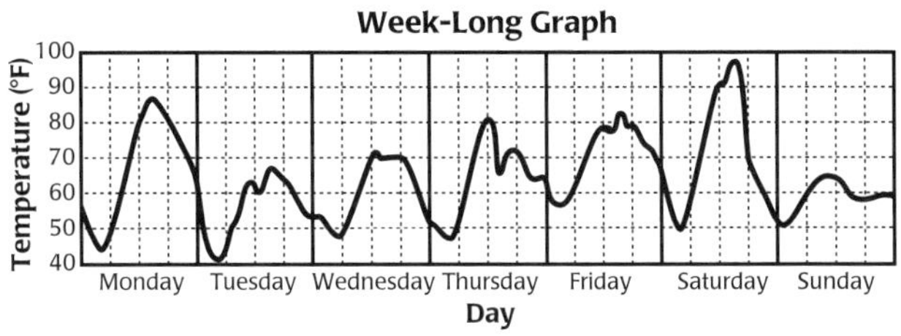

b. Yes, but not very accurately. You have to estimate the data rather than read from the graph.

c. Yes, but you cannot see exactly when it occurred.

d. Answers will vary. Sample responses:

Advantages:
A week-long graph has a convenient size. For example, you can put it in a notebook without having to fold it up.

You have a general view of the temperature during the whole week.

Disadvantages:
What you see is less accurate. Detail has been lost.

Hints and Comments

Materials Student Activity Sheet 12 (one per student)

Overview Students draw a compressed week-long graph and reflect on the graph's advantages and disadvantages.

Planning Have students work in pairs on problems **3a–c.** Students may work on problem **3d** individually. Problem **3d** can be used for informal assessment. This problem can also be assigned as homework. You may want to discuss problems **3b** and **3c** with the whole class before students continue with problem **3d.**

Comments about the Problems

3. a. It is important that students think about a strategy for drawing the graph accurately. Some reference points can be transferred easily: the high and low temperatures, the temperatures at 6:00 A.M., noon, 6:00 P.M., and midnight. The vertical dashed lines divide the day into periods of six hours. After some reference points are in, students can fill in the curve.

Some students will realize that six squares (or hours) from the day graph need to be compressed into a single square on the week-long graph.

b. It is not sufficient for students to note that the times have not been listed on the week-long graph.

d. Informal Assessment This problem assesses students' ability to understand the limitations of line graphs and to recognize the importance of scale in creating line graphs. You may want to assign this problem as homework.

Writing Opportunity You might suggest that students write their answers to problem **3d** in their journals.

4. **a.** Use your week-long graph from problem **3** to draw in the first week of the month-long graph on **Student Activity Sheet 12.**

   b. Finish the month-long graph using the weather descriptions given at the bottom of page 24.

Month-Long Graph

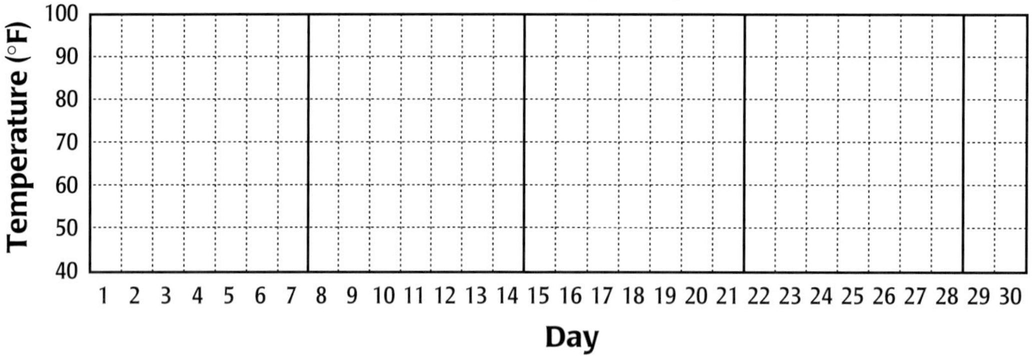

Months *into* Years

The month-long graph you made contains many ups and downs. Over 12 months, there could be 12 times as many ups and downs!

Suppose 12 month-long graphs are squeezed together to make a year-long graph.

Use the year-long graph on **Student Activity Sheet 12** to answer problem **5.**

Year-Long Graph

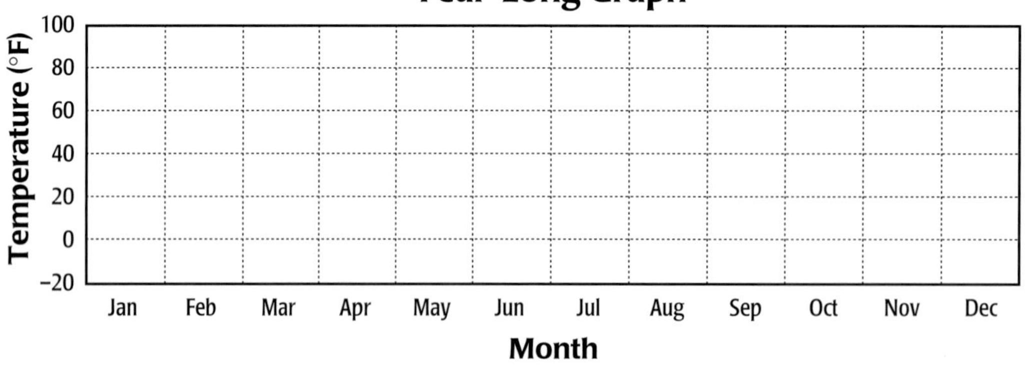

5. **a.** Notice that the temperature scale now goes from –20 to 100 degrees Fahrenheit. Why was the temperature scale changed for this graph?

   b. Sketch in the graph for the month of June.

   c. Sketch in what you think the graph might look like for the rest of the year.

4. a–b. Graphs will vary. Sample graph:

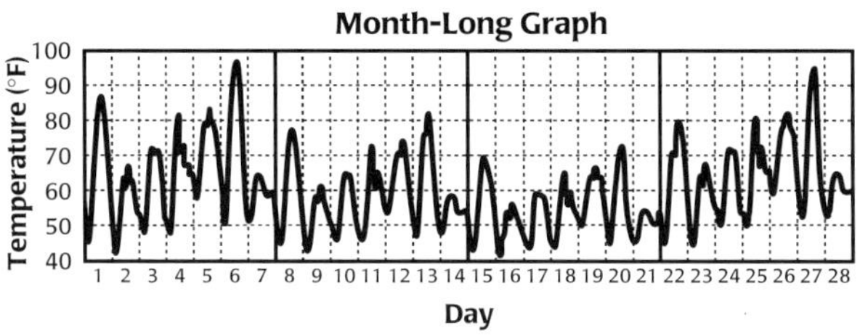

5. a. The temperature scale was changed to accomodate temperatures during cold months.

b.

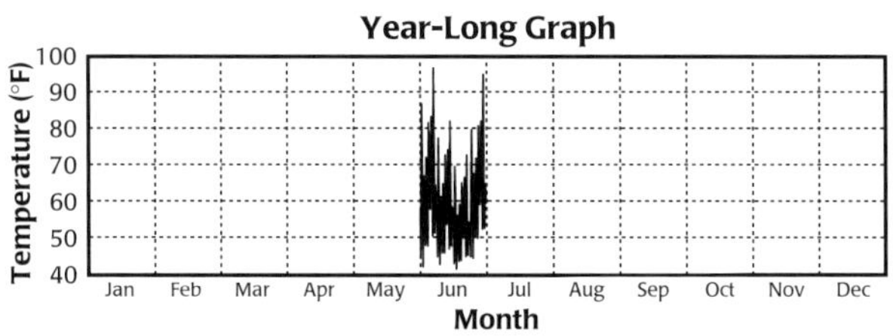

c. Graphs will vary. Sample graph:

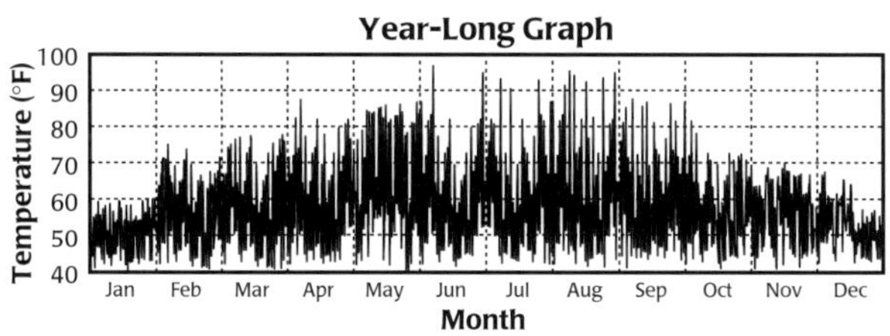

Hints and Comments

Materials transparency of year-long graph, optional (one per class); overhead projector, optional (one per class)

Overview Students use the week-long graph to draw the first week of the month-long graph. They use the weather descriptions to finish the month-long graph. Then they sketch a year-long graph.

About the Mathematics Drawing the compressed graphs will focus students' attention on the global features of the graphs, such as minimum and maximum values and variance. Students should realize that the longer the time period, the more the horizontal time scale is compressed, and the more detail is lost.

Planning Students may continue working in pairs on problems **4** and **5.** These problems may be assigned as homework.

You may want to make a transparency of the year-long graphs, as shown in the solution for problem **4c,** to show students how these graphs look when they are drawn in full detail.

Comments about the Problems

4. Homework This problem may be assigned as homework. Important features of these graphs include the number of ups and downs and the maximum and minimum temperatures for the day. Note that on the month-long graph shown in the solutions column, the daily variation is quite large—unlikely for most places. Also, students should draw their graphs in such a way that it is easy to see that the third week was the coldest.

5. Homework This problem may be assigned as homework.

b–c. The year-long graphs for these problems in the solutions column were generated by a computer. Student graphs need not be so detailed.

c. Answers may or may not reflect your local weather patterns. Students who have recently moved into your school district, in particular, may have diagrams that show other patterns. Accept all reasonable graphs. You might want to ask if the −20°F to 100°F temperature range is appropriate for your local weather patterns.

In the year-long graph, the ups and downs are squeezed together so much that you cannot see the line anymore.

Below you see a temperature graph for a year in a city in France. Note that the temperature scale is in degrees Celsius.

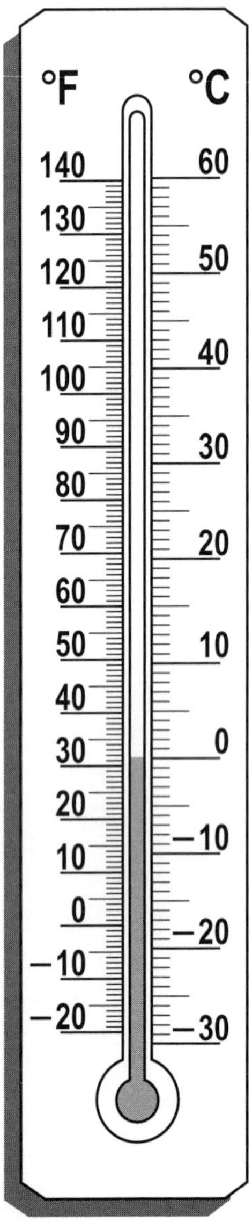

Year-Long Graph

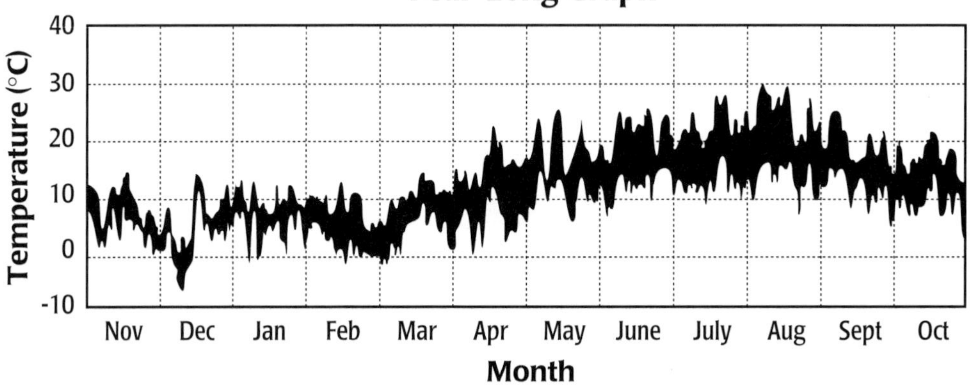

6. a. How is the graph above like your sketch for problem **5c?**

b. How is it different?

c. According to the graph, in which month was there the largest variation in temperature? How can you tell from the graph?

Use the graph above and the thermometer on the left to answer the next three questions in both Fahrenheit and Celsius temperatures.

7. What was the highest temperature in the French city that year, and when did it occur?

8. What was the lowest temperature, and when did it occur?

9. Water freezes when the temperature is 0°C or below. When did water freeze during that year?

Solutions and Samples
of student work

6. **a.** Answers will vary. Sample response:

   There are ups and downs for the changes in temperature.

   b. Answers will vary. Students may note that the line is thicker. You cannot see the individual ups and downs; you cannot read individual temperatures with any accuracy.

   c. April or August. April temperatures ranged from about 0°C to 23°C. August temperatures ranged from 7°C to 30°C. Explanations will vary. Students may say that you can tell by looking at the highest and lowest points in a month.

7. The highest temperature was 30°C, or 86°F. It happened twice, around August 8 and around August 16.

8. The lowest temperature was −7°C, or 19°F, around December 10th.

9. Water froze during December, at least two times in January, a few more times in February and in the beginning of March.

Hints and Comments

Overview Students investigate a temperature graph for a year in a city in France and look more closely at the global feature of variance.

About the Mathematics Most temperature graphs found in atlases are different from the year-long graph for a city in France in this section. The atlas graphs show mean, maximum, or minimum temperatures. From a mathematical point of view, these temperature line graphs are not continuous. The special attribute of the year-long graph in this section is that it looks like a string with a certain width but it actually has hundreds of ups and downs squeezed together.

Planning Students may work in pairs or individually on problems **6–9.** You may want to use problem **6** as informal assessment. When students finish these problems, discuss their answers to problem **6.**

Comments about the Problems

6. Informal Assessment This problem assesses students' ability to reason about global information such as regularity, maximum/minimum values, and variance as shown in line graphs.

7–8. You may want to discuss how to estimate when the highest and lowest temperatures occurred in a year-long map.

Summary

In this section, you looked at temperature graphs that showed temperature changes over different periods of time. To show longer periods of time, the graph had to be squeezed. The squeezed, or compressed, graph has advantages and disadvantages.

Advantages:
- The size of the graph is manageable.
- The graph gives a general view of how the temperature behaved.
- You can still see highs and lows.

Disadvantages:
- What you see is less accurate.
- If you compress a graph a lot, you can no longer read specific temperatures for specific days.

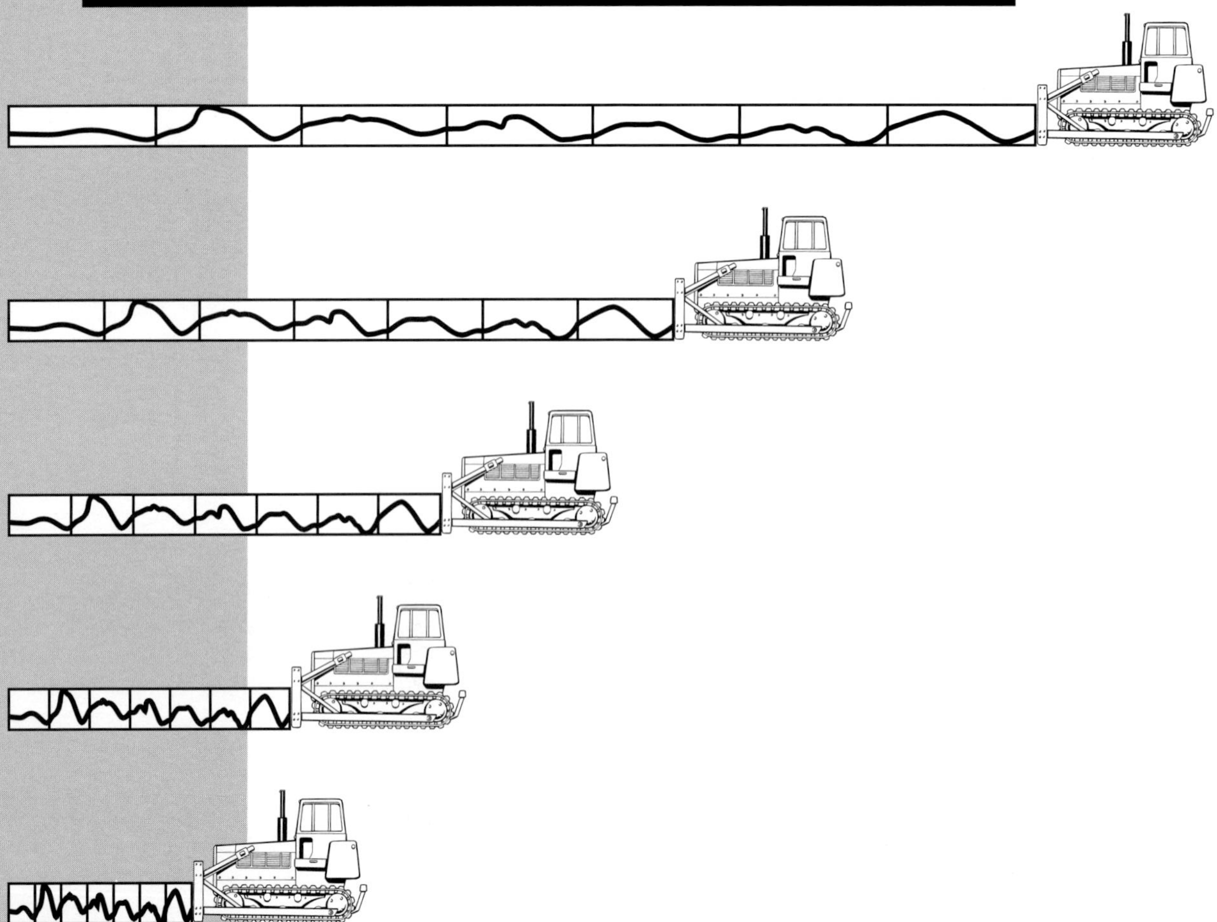

Solutions and Samples
of student work

Hints and Comments

Overview Students read and discuss the Summary, which highlights the advantages and disadvantages of squeezing a graph. There are no problems on this page for students to solve.

Planning After students have read the Summary, you may ask them to explain in their own words the advantages and disadvantages as presented. Students might be able to add some advantages and disadvantages. Then focus the discussion on the important features of all of these graphs, such as regularity, maximum and minimum values, slope, and variance.

Summary Questions

Below you see a temperature graph for one week. The labels for the axes have become separated from the graph, and the graph may not be right-side-up.

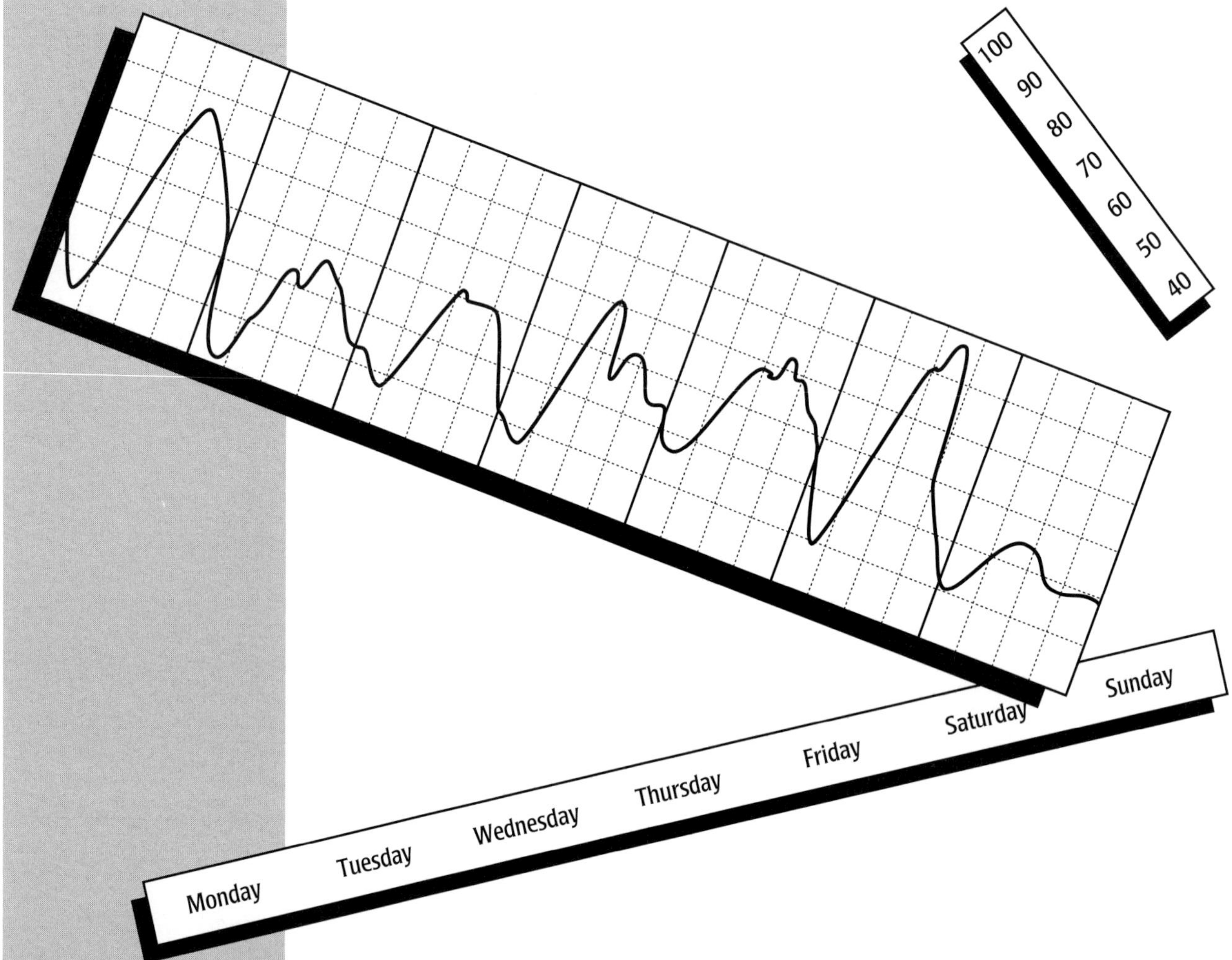

10. a. Which part of the graph is the bottom? How do you know?

b. Which day had the biggest change in temperature? The smallest?

c. Write one question that can be answered using this graph and one question that cannot be answered using this graph.

10. a. The graph is right-side up as shown. Explanations will vary. Sample response:

On most days, the high temperature for the day will be between noon and 6 P.M., and after 6 P.M., the temperature falls.

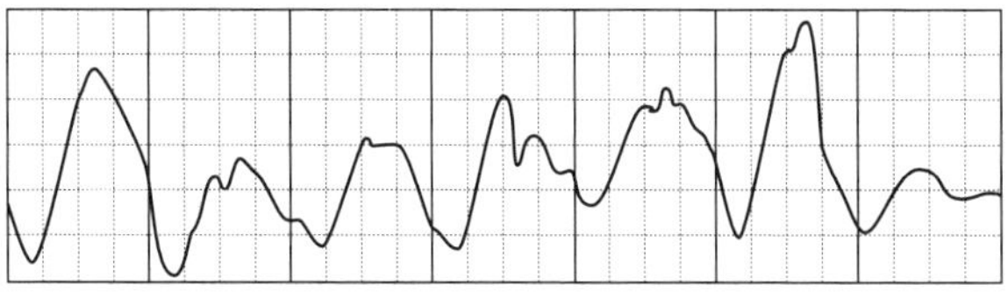

If you turn the page upside down, like this:

Upside Down

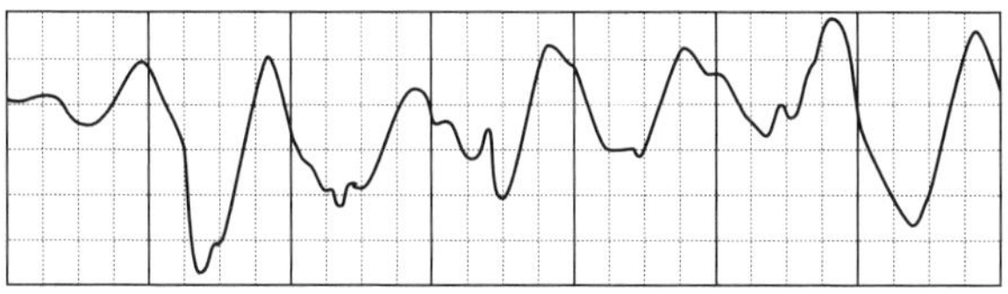

all the high temperatures would occur in the evening, which is unrealistic.

b. Saturday had the biggest change, almost 50° during the day. Sunday had the smallest change in temperature, only about 15°.

c. Questions will vary. Sample questions:

Questions that can be answered using this graph:
What was the low/high temperature for Thursday?
What day was the hottest?

Questions that cannot be answered using this graph:
What was the temperature at 4:30 P.M. on Tuesday?

Overview Students use what they learned about year-long graphs to solve the Summary problems.

Planning Students can work in pairs on problem **10.** Problem **10c** may be used for informal assessment. After students complete Section C, you may assign appropriate activities from the Try This! section, located on pages 39–42 of the *Tracking Graphs* Student Book, for homework.

Comments about the Problems

10. a. If students are having difficulty, ask them what they know about temperature changes throughout one day. You also might refer students to the seven day-long graphs at the beginning of Section C. This should help them realize that the temperature normally rises to a maximum temperature shortly after noon and then drops.

b. You might want to ask if the 50° range on Saturday is likely.

c. Informal Assessment Problem **10c** assesses students' ability to understand limitations of line graphs and to recognize the importance of scale in creating line graphs.

You may want to have students exchange questions and then solve each other's problems. Students would answer the first question using the graph. Ask students to explain why they cannot answer the second question using the graph.

Work Students Do

Students begin by investigating how speedometers show the speed of a car. They produce a line graph using the story of a car ride: One student holds a pencil on a piece of paper and moves the pencil up or down depending on the changes of the speed of the car, while another student pulls the paper at a steady rate. Students then create and examine continuous line graphs representing speed created from descriptions of various trips. Lastly, they create a line graph of a personal trip.

Goals

Students will:

- extract local, point-specific information from line graphs;
- create line graphs from numeric data;
- tell the story depicted by a line graph in terms of the context;*
- reason about global information such as regularity, maximum/minimum values, slope, and variance as shown in line graphs;
- create line graphs from verbal descriptions;
- recognize the importance of scale in creating line graphs;
- recognize the power of graphs for representing information and solving problems.**

* *This goal is assessed in other sections of the unit.*

** *This goal is introduced in this section and is assessed in other sections of the unit.*

Pacing

- approximately two 45-minute class sessions

About the Mathematics

In this section, a line graph is presented in a third way, as the record of a quantity—in this case, the speed of a car—changing over time. Reading the graph involves imagining the level of the speedometer at a certain point in the line graph. An important aspect of the mathematics of this technique is the realization of the need for a consistent scale. This way of constructing a line graph is in contrast to the "snapshot" approach, in which the graph is constructed from a discrete number of points, which are then connected. With the speed graphs, the data are continuous.

Materials

- paper or grid paper, page 71 of the Teacher Guide (two sheets per pair of students)
- colored pencils, page 71 of the Teacher Guide (three different colors per pair of students)
- tape, page 71 of the Teacher Guide, optional (one roll per pair of students)
- graph paper, page 75 of the Teacher Guide (one sheet per student)

Planning Instruction

You may wish to introduce the context of this section by discussing automobile speedometers. You might assign the following task as homework: Have students look at a car speedometer and make a drawing of it, including the scale. They could bring their drawings to class and discuss the similarities and differences between speedometers. You might want to discuss speeds indicated in miles per hour and in kilometers per hour; however, in this section, all speeds are given in miles per hour.

Students can work on problems 1–4 in pairs. Problems 5 and 6 can be worked on in pairs or individually. Have students work on problems 7 and 8 individually. Students may return to working in pairs or continue to work individually on problem 9.

There are no optional problems in this section.

Homework

Problem 9 (page 74 of the Teacher Guide) can be assigned as homework. The Extension (page 73 of the Teacher Guide) can also be assigned as homework. After students complete Section D, you may assign appropriate activities from the Try This! section, located on pages 39–42 of the *Tracking Graphs* Student Book. The Try This! activities reinforce the key mathematical concepts introduced in this section.

Planning Assessment

- Problem 6 can be used to informally assess students' ability to create line graphs from numeric data.
- Problems 7 and 8 can be used to informally assess students' ability to extract local, point-specific information from line graphs and to reason about global information such as regularity, maximum/minimum values, slope, and variance as shown in line graphs.
- Problem 9 can be used to informally assess students' ability to create line graphs from verbal descriptions and to recognize the importance of scale in creating line graphs.

D. GRAPHS DEPICTING SPEED

Making SPEED Graphs

On the dashboard of every car, there is a speedometer. Most of them have a needle that moves along a row of numbers that tells how fast the car is moving. For example, if the needle is on the number 50, the car is moving at a speed of 50 miles per hour (mph).

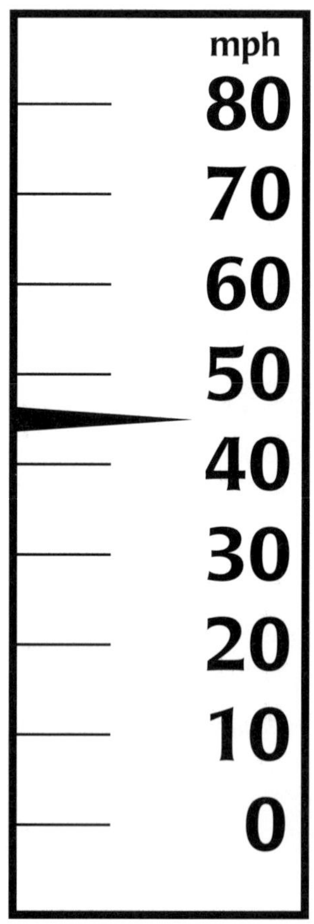

1. What is the speed of a car, in miles per hour, if its speedometer looks like the one on the right?

2. Point to where the needle of the speedometer would be for each of the following situations:

   a. The car is standing still.

   b. The car is moving at a speed of 30 miles per hour.

3. **a.** What happens to the needle if the car speeds up?

   b. What happens to the needle if the car slows down?

   c. What is the car doing if the needle stays in the same position for a while?

   d. What happens to the needle if the car stops suddenly?

Solutions and Samples
of student work

1. about 45 miles per hour

2. a–b.

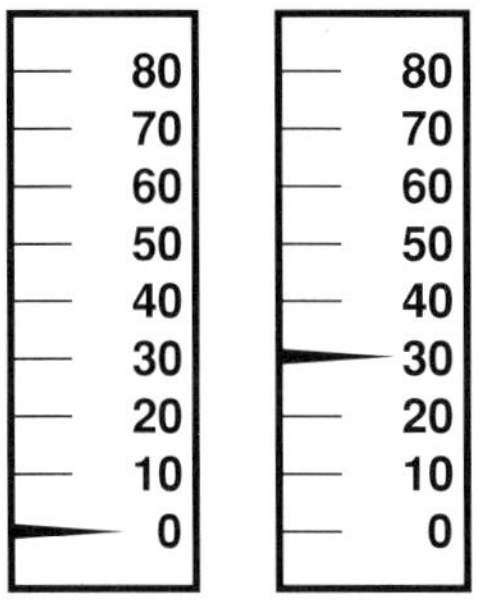

3. a. It moves upward.

   b. It moves downward.

   c. The car is moving at a constant speed (or, if the needle is at 0, the car is standing still).

   d. It moves quickly to zero and stays there.

Hints and Comments

Overview Students investigate how the needle of a speedometer indicates the speed of the car when the car is moving at different speeds.

Planning Have students work in pairs on problems **1–3.** Discuss problem **3** before students continue working on Student Book page 30.

Comments about the Problems

3. Make sure students understand this problem before they continue with the activity on the next Student Book page. You might provide additional practice by having students imitate the movement of the needle with their hand as you describe the movements of a car. For example, you might say, "The car starts moving from the parking place; then it stops. It turns onto a residential street and reaches a speed of 30 mph. It moves up the on-ramp of a highway and speeds up to 50 mph. . . ."

Activity

Pair up with another student to make a graph of several car trips. For each line on the graph, one student will draw, while the other student pulls the paper to the left. The student who draws should move the pencil up when the car goes faster and down when the car goes slower.

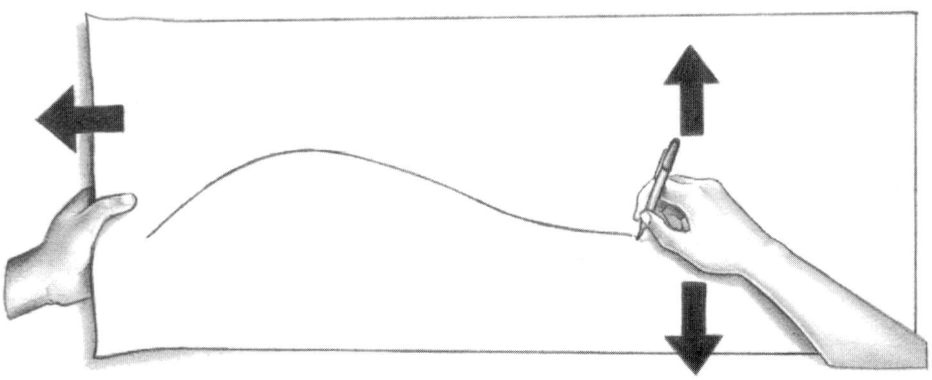

Record the following trips on the same graph. Use a different color for each trip.

4. a. From a stop, gradually speed up to 55 miles per hour and stay at that speed until you reach the end of the paper.

b. From a stop, quickly speed up to 40 miles per hour, then gradually slow down to a stop.

c. Starting at 55 miles per hour, slow down and then speed up again. Do this repeatedly.

4. a–c. Graphs will vary. Sample graph:

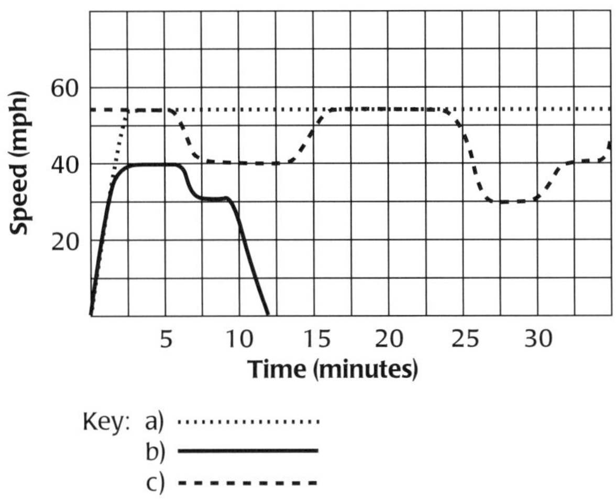

Materials colored pencils (three different colors per pair of students); paper or grid paper, (two sheets per pair of students); tape, optional (one roll per pair of students)

Overview Students create line graphs using the story of a car ride. One student puts a pencil to paper and moves it up and down in imitation of a speedometer needle, while another student pulls the drawing paper to the left.

About the Mathematics In this context, a line graph is presented as the record of a quantity (speed) changing over time. This way of constructing a line graph is in contrast to the "snapshot" approach of Section A, in which the graph is constructed from a discrete number of points that are then connected. The data in the speed graphs are continuous. This activity helps students recognize the need for and purpose of scales on graphs.

Planning Have students work in pairs on problem **4.** You may want to have students tape two sheets of paper together in order to get a longer sheet to make the graph. When students have finished this activity, discuss their results with the whole class. Suggestions for a class discussion are mentioned below.

Comments about the Problems

4. Students should try to pull the paper at a constant speed. This graph intentionally uses no scale so that there will be variability in the responses. Students should compare graphs for readability and consistency.

You may wish to ask students the following questions:

- *Why does the graph of one pair of students look different from that of another pair?* [Perhaps the graphs use different scales and were made with different pulling rates. This can be related to the compressed temperature graphs from Section C.]
- *What does pulling the paper represent?* [the passage of time]
- *What happens if you pull the paper faster? slower?* [Time speeds up/slows down, or in other words, the time scale becomes more or less compressed.]
- *Looking at the three graphs on the single page, what does it mean when the graphs intersect?* [The speeds are the same. Some students may make the mistake of thinking that the cars are passing or hitting one another.]

Reading *SPEED* Graphs

Below you see the speed graph for a car ride.

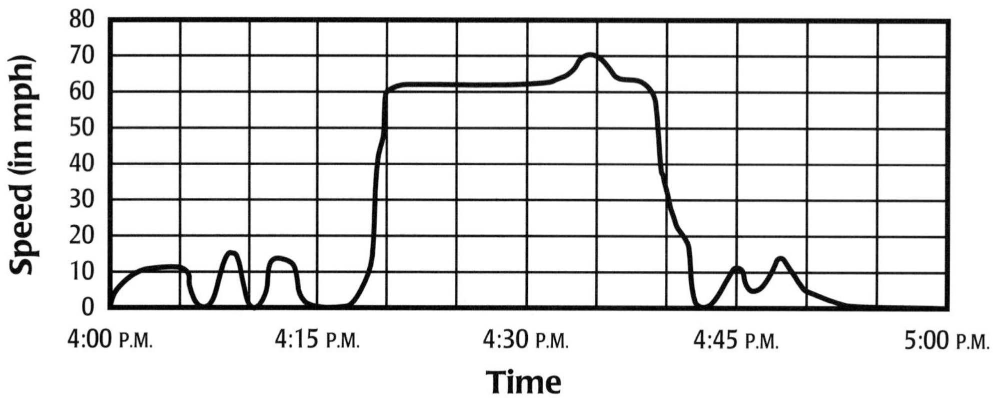

5. **a.** What might have been happening during the first 15 minutes of the ride?

b. How often did the car stop during this period?

6. What type of road could the car have been on between 4:20 P.M. and 4:30 P.M.?

7. The line graph is almost horizontal between 4:20 P.M. and 4:30 P.M. What can you say about the speed during this period?

8. Suppose the speed limit on that road is 65 miles per hour. Did the car exceed the speed limit during the ride?

Solutions and Samples
of student work

5. a. Answers will vary. Sample response:

They drove slowly and stopped a couple of times. Maybe there was a lot of traffic, or they drove out of a parking lot. The last stop lasted a couple of minutes.

b. The car stopped three times.

6. Answers will vary. Some students may say that it was probably an interstate highway or a country road with little traffic, because they were able to go quite fast.

7. The speed is constant, at around 62 mph.

8. Yes, the car traveled at a speed of 70 mph around 4:35 P.M. Some students may add that perhaps it was necessary to pass some other cars.

Hints and Comments

Overview Students interpret a given speed graph of a car ride.

About the Mathematics Reading the graph involves imagining the level of the speedometer at a certain point in the line graph.

Planning Students may work in pairs or individually on problems **5** and **6.** Discuss these problems with the whole class and have students continue working individually on problems **7** and **8.** Problems **6–8** can be used as informal assessment.

Comments about the Problems

5. Students should be able to translate the first part of the line graph in terms of the context. If students still confuse speed and distance, you may refer to problem **3** of this section and (re)do the suggested additional activity.

6. Informal Assessment This problem assesses students' ability to create line graphs from numeric data.

7–8. Informal Assessment These problems assess students' ability to extract local, point-specific information from line graphs and to reason about global information such as regularity, maximum/minimum values, slope, and variance as shown in line graphs. When discussing problems **7** and **8** with the whole class, you may ask students the following questions:

- *Can you find the distance the car traveled between 4:20 and 4:30?* [Ten minutes is about $\frac{1}{6}$ of an hour, and the car traveled at a speed of about 61 mph, so in 10 minutes, it traveled $\frac{1}{6}$ of 61 miles, or about 10 miles.]
- *Why can't you find the distance the car traveled between 4:15 and 4:20?* [Because the speed is changing in that period]

Extension You may wish to ask students to make graphs of 24-hour driving days for drivers whose jobs depend on driving—for example, long-distance truck drivers, pizza delivery drivers, or taxi drivers. Each graph should reflect the particularities of the job. For instance, the truck driver graph should show long periods of driving at highway speeds, while a pizza delivery graph would show many stops for dropping off pizzas.

Summary

Graphs can show the speed of a moving object, such as a car, over time.

The speed scale along the side of a graph and the time scale along the bottom are both important for interpreting a speed graph.

Summary Questions

Activity

9. Take notes about your trip to school on a normal day. You can do this for whatever type of transportation you use: walking, a bike, a school bus, a car, or a subway. You should record your starting time and the times of events like stops for traffic lights, waiting at the bus stop, stops to pick up other kids, and so on.

Record the trip on a graph like the one on the previous page. Estimate your speed during the times you are moving. You should also supply a written description of your trip. Be prepared to have someone evaluate your graph by comparing it to your written description.

9. Graphs and descriptions will vary. Some features to look for in student responses are as follows:
   - The speed should be zero at the beginning and end of the trip.
   - Most trips will have several stops that should be shown with the line graph meeting the horizontal axis.
   - Each axis should be labeled in units of equal intervals.

Materials graph paper (one sheet per student)

Overview Students read the Summary, which reviews the importance of using a scale to interpret a speed graph. They then create a speed graph of a personal trip.

Planning Have students read and discuss the Summary. Then discuss strategies for recording their trips before they begin working on problem **9.** For example, students might want to pay particular attention to the times that they are stopped. This activity can be done in pairs if students travel to school together. You can also assign this problem partially or totally as homework. After students complete Section D, you may assign appropriate activities from the Try This! section, located on pages 39–42 of the *Tracking Graphs* Student Book, for homework.

Comments about the Problems

9. **Informal Assessment** This problem assesses students' ability to create line graphs from verbal descriptions and their ability to recognize the importance of scale in creating line graphs. It may be assigned as homework.

   Since this is one of the first times students are asked to make their own speed graphs, they may find this a difficult task. In particular, creating the scale is new and should not be rushed. It may be helpful to discuss different speeds of walking, cycling, and so on, before students start to work.

   As an alternative, students can make up a trip and draw a graph that would represent it.

Work Students Do

Students investigate graphs of tides. Students use these graphs to solve problems about ships that want to visit a harbor. They investigate various situations related to the amount of water ships "draw," that is, the depth of water necessary to support the ships. They also examine how the tides must be taken into account when a ship enters or leaves a harbor.

Goals

Students will:

- extract local, point-specific information from line graphs;
- create line graphs from numeric data;
- tell the story depicted by a line graph in terms of the context;
- interpolate data points to fill in line graphs when appropriate;
- reason about global information such as regularity, maximum/minimum values, slope, and variance as shown in line graphs;
- recognize the power of graphs for representing information and solving problems;
- recognize the importance of scale in creating line graphs.*

** This goal is assessed in other sections of the unit.*

Pacing

- approximately two 45-minute class sessions

About the Mathematics

This section reinforces the mathematical concepts in previous sections regarding interpreting line graphs. The line graphs in this section are all tidal graphs, which are not only continuous, but also periodic. The concept of periodicity is made explicit in the grade 7/8 unit *Ups and Downs.* Students will consider the shape of these graphs as well as relevant maximum and minimum points related to the rising and lowering of the water level. A limitation on water depth, which plays an important role in the problems about ships entering the harbor, can be shown by drawing a horizontal line across the entire graph at that depth.

In this section, students will apply another transformation of graphs: a translation. The translations that will be discussed arise in a natural way. If the tides at two harbors change in the same way and there is a constant difference between the water heights at the two harbors, then the tidal graph of one harbor can be used to make the tidal graph of the other harbor. Both graphs will have the same shape, but the graph of one harbor will start at a higher level than the graph of the other harbor.

This section emphasizes the power of graphs for representing information and solving problems.

Materials

- Student Activity Sheets 13 and 14, pages 125 and 126 of the Teacher Guide (one per student)
- transparency of Student Activity Sheet 13, page 81 of the Teacher Guide, optional (one per class)
- overhead projector, pages 81, 85, and 87 of the Teacher Guide, optional (one per class)
- colored pencils, page 83 of the Teacher Guide, optional (a few per group of students)
- transparencies of the city's harbor tidal graph and of that same graph without axes, page 85 of the Teacher Guide, optional (one of each per class)
- transparency of the tidal graph of the plant's harbor, page 87 of the Teacher Guide, optional (one per class)

Planning Instruction

You may wish to introduce the context of this section by discussing tides. Some students will be more familiar with tides than others. Ask students about how many times there are high tides and low tides during a 24-hour period (two high and two low). If you live near water that has tidal fluctuations, you may obtain local times for high and low tides. You can then focus the discussion on water depths (measured in meters), or perhaps on different types of ships and how their shape and size affect how much water they draw.

You may choose to have students work individually on problem 6, and individually or in pairs on problems 9 and 10. Students may work on the remaining problems in pairs or small groups.

There are no optional problems in this section.

Homework

Problems 2 (page 80 of the Teacher Guide), and 4 and 5 (page 82 of the Teacher Guide) can be assigned as homework. In addition, the Extension activity (page 89 of the Teacher Guide) and the Writing Opportunity (page 89 of the Teacher Guide) may be assigned as homework. After students complete Section E, you may assign appropriate activities from the Try This! section, located on pages 39–42 of the *Tracking Graphs* Student Book. The Try This! activities reinforce the key mathematical concepts introduced in this section.

Planning Assessment

- Problem 2 can be used to informally assess students' ability to reason about global information such as regularity, maximum/minimum values, slope, and variance as shown in line graphs. It can also be used to assess their ability to recognize the power of graphs for representing information and solving problems.
- Problem 6 can be used to informally assess students' ability to create line graphs from numeric data and to interpolate data points to fill in line graphs when appropriate.
- Problem 9 can be used to informally assess students' ability to extract both local, point-specific information from a line graph and to recognize the power of graphs for representing information and solving problems.
- Problem 10 can be used to informally assess students' ability to tell the story depicted by a line graph in terms of the context.

E. GRAPHS OF TIDES

CITY *Harbor*

If you visit an ocean beach, you may notice that the water level can vary. At low tide, the sea is further away, and the beach is wide. At high tide, the sea is closer, and the beach is narrow.

Most coastal cities have harbors, which are areas of the sea that have land on three sides. The land protects ships from stormy seas. The captains of ships entering harbors have to know if it is high or low tide to determine whether the water is deep enough to keep the ship afloat. Harbors are sometimes dredged, or dug deeper, so that ships can stay in the harbor even during low tide.

Below and on **Student Activity Sheet 13** is the tidal graph of a harbor for one summer day. Vertically, you see the water depth in meters.

1. a. At what times was the water level the highest? How high did it go?

b. At what times was the water level the lowest? How low was it?

c. What was the range of the water's depth?

d. At what times was the water 4 meters deep?

e. Between what times was the water at least 6 meters deep?

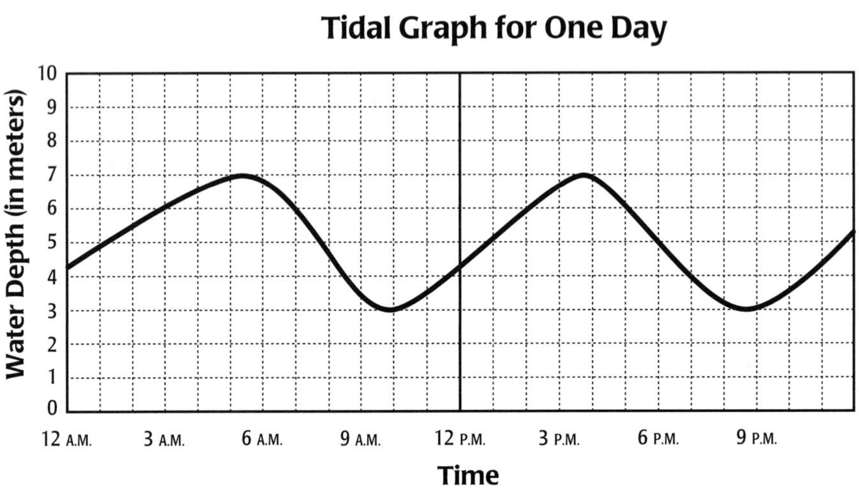

Solutions and Samples
of student work

1. **a.** Around 5:30 A.M. and 3:45 P.M., the water level was at 7 meters.

   b. Around 9:45 A.M. and about 8:30 P.M., the water level was at 3 meters.

   c. The water depth was between 3 and 7 meters.

   d. at approximately 8:30 A.M., 11:30 A.M., 7:00 P.M., and 10:30 P.M.

   e. Approximately from 3:00 A.M. until 7:00 A.M., and from 2:00 P.M. to 5:00 P.M. Some students may show this by pointing to the range on the graph:

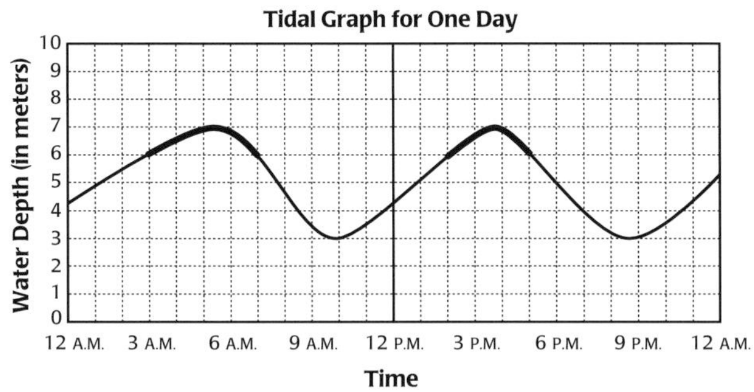

Hints and Comments

Materials Student Activity Sheet 13 (one per student)

Overview Students investigate a tidal graph that shows the water depth in a harbor during a whole day.

About the Mathematics This section reinforces the mathematical concepts of the previous sections. Continuous graphs are investigated for maximum and minimum points as well as a specific range of values. Because tides are periodic events, the graphs show a regular, repetitive pattern. Students will further investigate periodic graphs in the grade 7/8 unit *Ups and Downs.*

Planning Students can work in small groups or in pairs on problem **1.**

Comments about the Problems

1. You may want to use this problem to check students' abilities to extract local information from line graphs.

   c. If students are having difficulty with the word *range,* you may want to provide a simple definition.

   e. If students have difficulty with this problem, you may suggest that they first color all parts of the graph that show a water depth of six meters or more. Then students can use the colored parts to look for the time periods that belong to each colored part.

This big ship "draws" 5 meters. That means that the underside of it is 5 meters below sea level, so the ship needs water that is at least 5 meters deep.

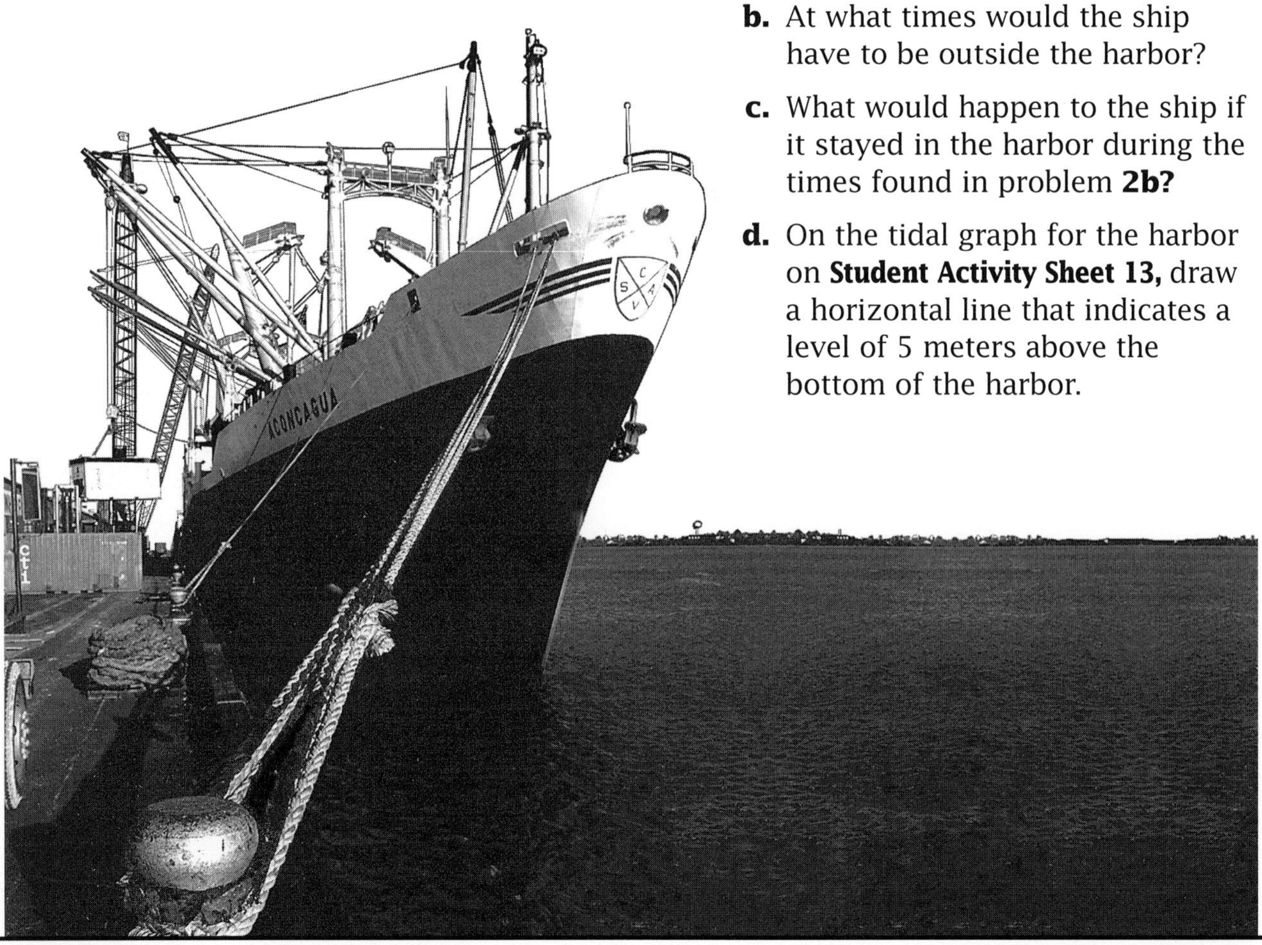

2. a. Explain why the above ship could not spend a whole day in the harbor you studied on the previous page.

b. At what times would the ship have to be outside the harbor?

c. What would happen to the ship if it stayed in the harbor during the times found in problem **2b?**

d. On the tidal graph for the harbor on **Student Activity Sheet 13,** draw a horizontal line that indicates a level of 5 meters above the bottom of the harbor.

Solutions and Samples
of student work

2. **a.** The ship needs a water depth of at least 5 meters, and the graph shows that at certain times during the day, the water depth in the harbor can be less than 5 meters.

   b. The ship has to be outside the harbor when the depth is less than 5 meters. This happens from approximately 12:00 A.M. to 1:00 A.M., from 7:45 A.M. to 1:00 P.M., and from 6:00 P.M. to 11:45 P.M.

   c. The ship would drop with the water level and run aground (touch the bottom of the harbor).

   d.

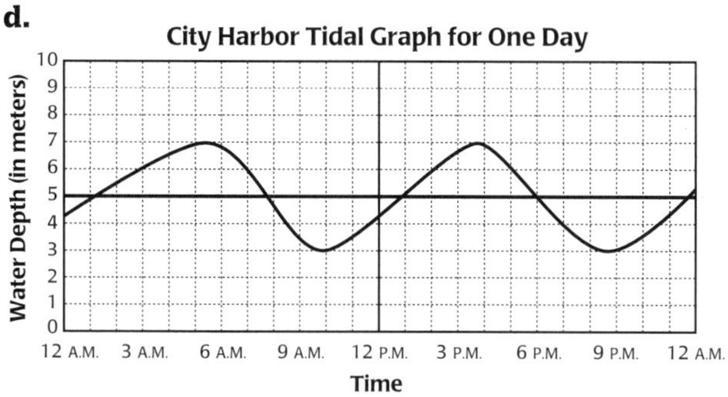

Hints and Comments

Materials transparency of Student Activity Sheet 13, optional (one per class); overhead projector, optional (one per class)

Overview Students use the graph to solve problems about a ship that can float only if the water is at least 5 meters deep.

Planning You may want to have students work on problem **2** in small groups and then discuss this problem with the whole class. Ask students how they organized their work in order to get their answers. Problem **2** can be used as informal assessment and may be assigned as homework.

Comments about the Problems

2. **Informal Assessment** This problem assesses students' ability to reason about global information such as regularity, maximum/minimum values, slope, and variance as shown in line graphs. It also assesses their ability to recognize the power of graphs for representing information and solving problems. You may want to assign this problem as homework.

   b. To solve this problem, students must first realize that the ship has to be outside the harbor when the water depth is 5 meters or less. Then they can look at the graph to learn the times when this occurs.

   If students have difficulty with this problem, you may ask them to do problem **2d** first and then use its answer to solve problem **2b.**

   When discussing this problem, it might be useful to indicate on a transparency of Student Activity Sheet 13 the part of the graph that shows a depth of at least 5 meters and then color the parts of the graph that show a depth of less than 5 meters (see the sample solution for problem **2d**).

   c. Ships are not meant to rest on the harbor floor. Because of the shape of the hull, most ships would fall to one side and tip over.

   d. You may want to have students explain how problems **2d** and **2b** are related, or how the answer to problem **2d** can be used to solve problem **2b.**

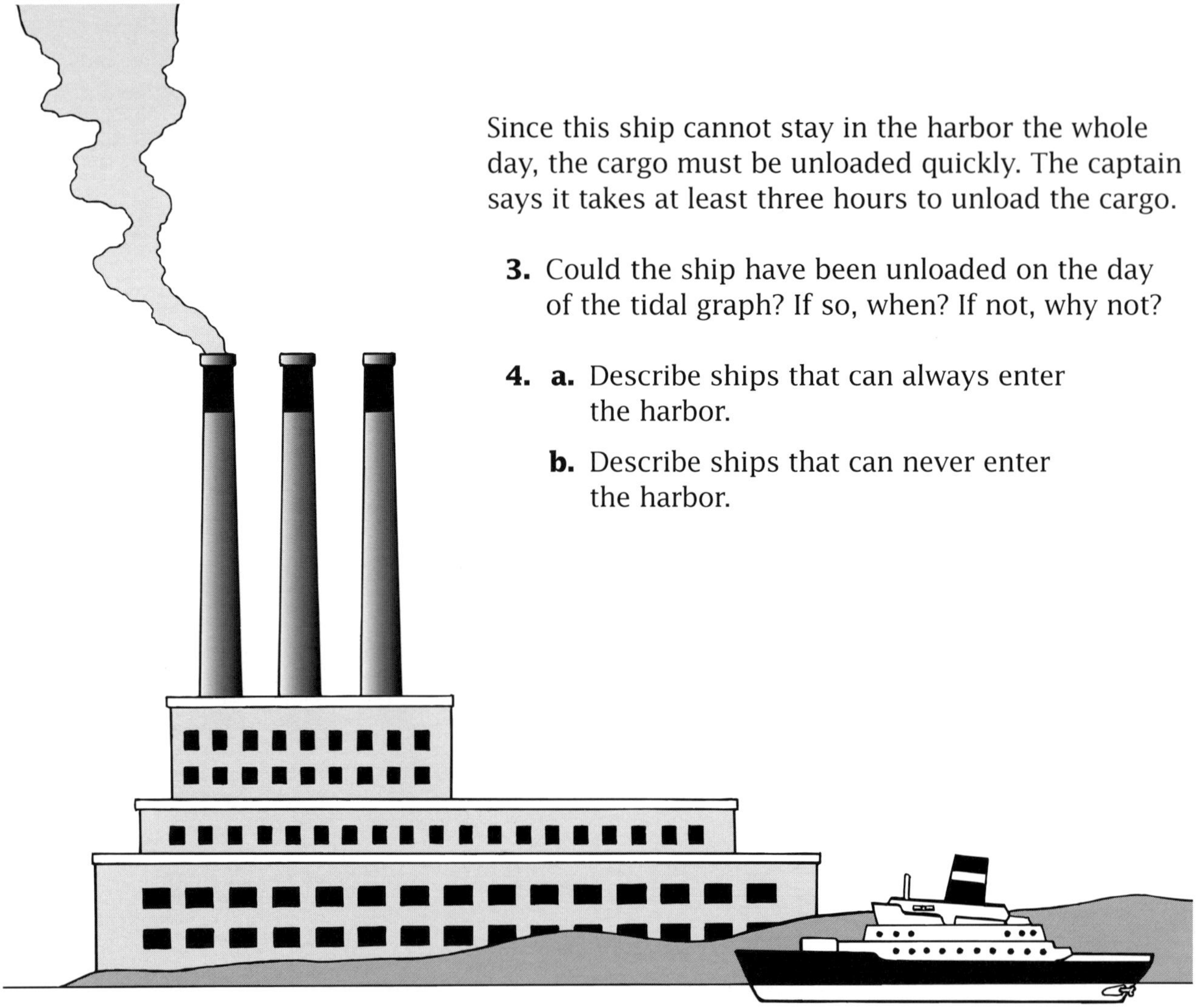

Since this ship cannot stay in the harbor the whole day, the cargo must be unloaded quickly. The captain says it takes at least three hours to unload the cargo.

3. Could the ship have been unloaded on the day of the tidal graph? If so, when? If not, why not?

4. **a.** Describe ships that can always enter the harbor.

b. Describe ships that can never enter the harbor.

PLANT *Harbor*

Outside the city is a manufacturing plant. This plant is located on the coastline and has its own harbor. During low tide, the plant's harbor is only 1 meter deep. Its tide moves up and down by the same amount and is on the same schedule as the tide of the city's harbor.

5. **a.** How deep is the water in the plant's harbor during high tide?

b. When is the water in the plant's harbor at least 4 meters deep?

Solutions and Samples

of student work

3. Yes. Times and explanations will vary. Sample responses:

The ship can be in the harbor between 1:00 A.M. and 7:45 A.M. It takes three hours to unload, so they have to start unloading at least three hours before 7:45 A.M., that is, at 4:45 A.M.

One possibility is entering the harbor between 1:00 P.M. and 3:00 P.M. They have to start unloading before 3:00 P.M. in order to leave the harbor just before 6:00 P.M.

4. a. ships that draw less than 3 meters

b. ships that draw more than 7 meters

5. a. 5 meters. Strategies will vary. Sample strategies:

Strategy 1

At low tide, the plant harbor is 1 meter deep. At the same time, the city harbor is 3 meters deep. So the plant harbor is 2 meters shallower than the city harbor, and this will be true at all times. Therefore, at high tide the water depth in the city harbor is 7 meters and in the plant's harbor it is 7 − 2 = 5 meters.

Strategy 2

The difference between high tide and low tide is 4 meters. This counts not only for the city harbor, but also for the plant's harbor. So when the plant's harbor is 1 meter deep at low tide, the depth will be 1 + 4 = 5 meters at high tide.

b. Between 3:00 A.M. and 7:00 A.M. and again between 2:00 P.M. and 5:00 P.M. It happens when the water in the city harbor is at least 6 meters deep.

Hints and Comments

Materials colored pencils, optional (a few per group of students)

Overview Students solve two more problems about the ship and the city harbor. They then start to investigate what happens in the plant's harbor, where the tide moves up and down by the same amount as in the city harbor, but the depth at low tide is only 1 meter.

About the Mathematics In Section C, students investigated what happens to a graph when the horizontal axis is compressed. In problem **5,** students start to investigate another operation on graphs: a vertical translation of a graph.

Planning Students can work on problems **3–5** in pairs or small groups. Discuss problems **3** and **5** with the whole class before students continue working on the problems on the next page. Problems **4** and **5** may be assigned as homework.

Comments about the Problems

3. If students are having difficulty, you may want to encourage them to use the graph and a colored pencil to indicate the suitable depths and times.

Students may use their answers to problem **2b** when working on this problem.

4–5. Homework These problems can be assigned as homework.

4. a. Students might answer with a specific type of boat, such as a canoe. They should be able to justify their answer in terms of how much the boat draws.

5. a. Discuss students' strategies as a whole class.

b. For this problem, only the first sample strategy given for problem **5a** can be used. Note that this problem is related to problem **1e.**

Look at the grid for the plant's harbor on **Student Activity Sheet 13.** Notice that it contains two points of the harbor's tidal graph.

Plant Harbor—Tidal Graph for One Day

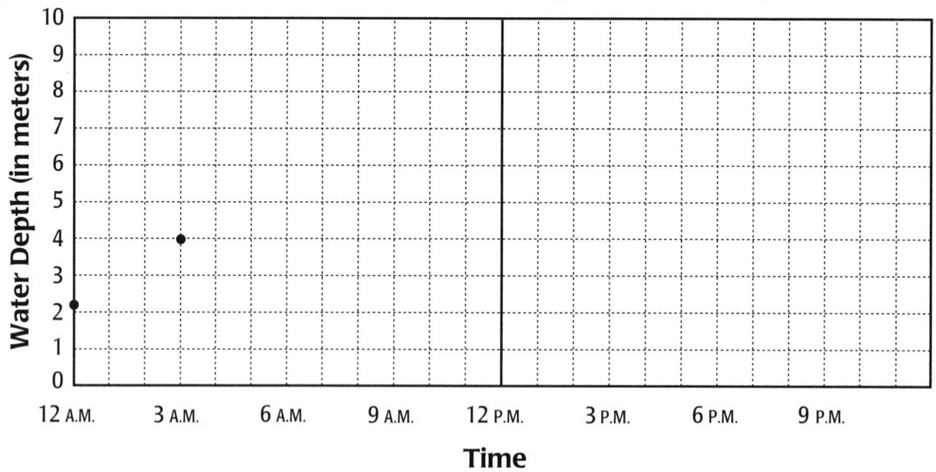

The tidal graph for the plant's harbor can be found using the graph for the city harbor.

6. a. Find the water depths every three hours and plot them on the graph on **Student Activity Sheet 13.**

b. Finish the graph by connecting these points with a smooth curve.

c. Compare the plots of the water depths for the plant and city harbors.

Solutions and Samples
of student work

6. a–b.

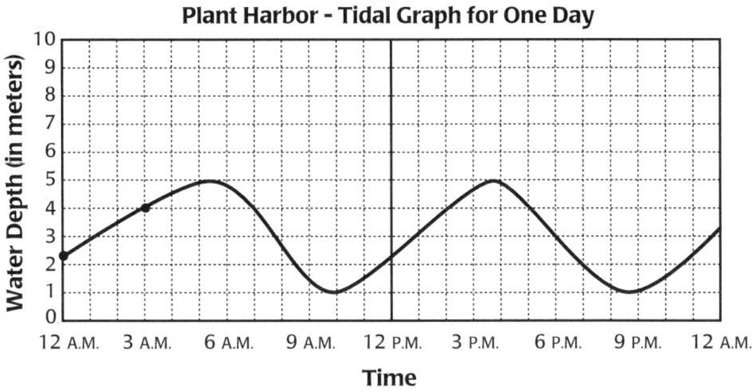

c. Comparisons will vary. Students may note that the shape is exactly the same; the graph is just lower.

Hints and Comments

Materials Student Activity Sheet 13 (one per student); transparencies of the city harbor tidal graph and of that same graph without axes, optional (one of each per class); overhead projector, optional (one per class)

Overview Students use the tidal graph for the city harbor to construct the tidal graph for the plant's harbor.

Planning You might want students to work individually on problem **6.** Problem **6** may be used as an informal assessment.

Comments about the Problems

6. Informal Assessment This problem assesses students' ability to create line graphs from numeric data and verbal descriptions. It also assesses their ability to interpolate data points to fill in line graphs when appropriate.

c. You might make a transparency of the city harbor's tidal graph (on Student Activity Sheet 13) as well as a transparency of that graph without the axes. The second transparency can be superimposed on the first and then moved down to show how the plant harbor graph looks relative to that for the city harbor.

Ask students how the graph of the city harbor can be changed to become that of the plant harbor. Students might suggest moving the graph down. Some students may suggest renumbering the vertical axis.

Suppose that on the day represented by the tidal graph on page 36, a ship is on the open sea. It contains coal for the manufacturing plant. The captain wants to unload the cargo in the plant's harbor. The ship draws 2.5 meters. After unloading, the ship will draw 2 meters. At 1:30 P.M., the captain announces that the ship can reach the harbor in 2 hours.

7. Will the ship reach the plant harbor in time to unload if unloading takes one hour?

It is a very hot, humid day. When the ship arrives, the captain decides to wait until after dark to enter the harbor. It will be cooler then.

8. In this case, when should the captain plan to enter the harbor?

Summary

Some graphs, such as graphs of tides, go up and down in a regular way. This pattern is different from graphs that go up and down irregularly, such as the graphs of car speeds.

Graphs that go up and down in a regular way can have the same shape but begin at different places on the vertical axis, as the graphs for the city and plant harbor tides did.

Solutions and Samples
of student work

7. Yes. Explanations will vary. Sample response: The ship can reach the harbor by 3:30 P.M. and be emptied by 4:30 P.M. At that time the water level is still above 2 meters.

According to the graph, even if the ship needs two hours to get out of the harbor, the water will still be higher than than 2 meters:

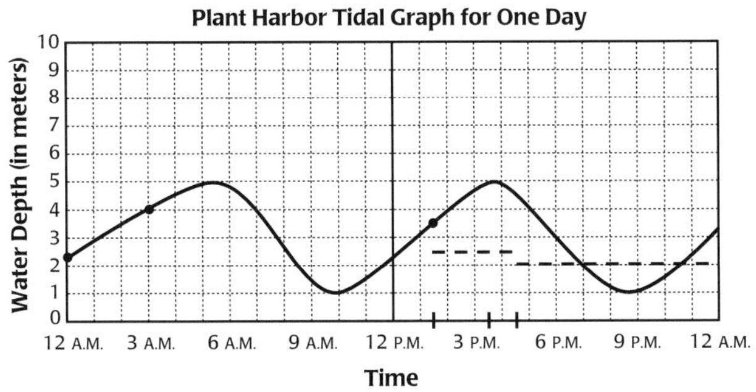

8. After 11:30 P.M. Explanations will vary. Sample response:

They can enter the bay again when the water becomes deeper than 2.5 meters. This will happen around 11:30 P.M.

Hints and Comments

Materials transparency of the tidal graph of the plant's harbor, optional (one per class); overhead projector, optional (one per class)

Overview Students use the tidal graph to solve two problems about a ship that must enter the harbor.

Planning You may have students work in pairs or small groups to solve problems **7** and **8.** For some students these problems may be a challenge, so you might give them extra credit for solving the problems correctly.

A copy of the tidal graph of the plant's harbor on a transparency (as shown in the Solutions column for problem **7**) may be useful for a class discussion of these problems.

Comments about the Problems

7–8. Students have to read the information very carefully. Encourage them to use the graph and the time scale to figure out whether or not it is possible to leave the harbor on time. They should use the tidal graph of the plant's harbor that they made in problem **6.**

You may want to ask one group of students to work with a copy of the graph on a transparency, so that they can show their strategy and solution to the whole class.

Summary Questions

Below and on **Student Activity Sheet 14,** you see the tidal graph for a harbor on a certain day. Someone has relabeled the vertical axis by hand to show the tide of a nearby harbor.

Tidal Graph for One Day

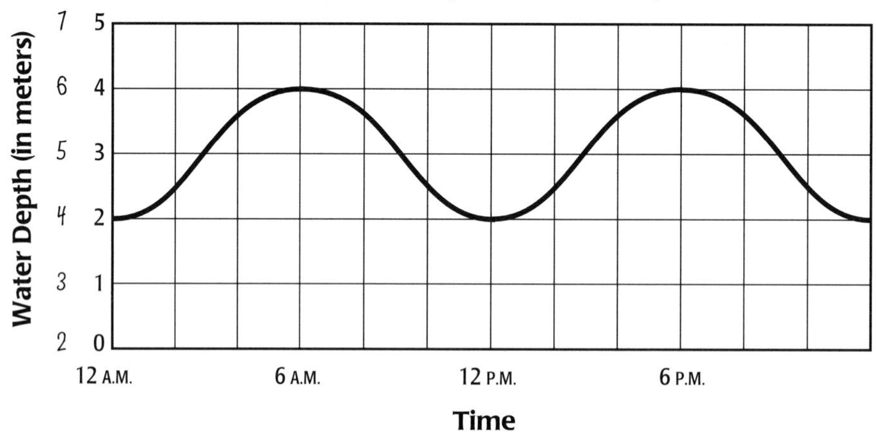

Use the graph on **Student Activity Sheet 14** to answer the following questions:

9. Consider the harbor for which the graph was originally made.

a. At what time is low tide? How low is the tide?

b. At what time is high tide? How high is the tide?

c. When is the water in the harbor rising? Indicate the answer by coloring the tidal graph on **Student Activity Sheet 14.**

d. A ship draws 3.5 meters. It enters this harbor in the afternoon. By what time must this ship leave the harbor?

10. Now consider the nearby harbor and use the handwritten labels on the vertical axis. Is the water in the nearby harbor deeper or more shallow than in the other harbor? How much deeper or shallower is it?

9. **a.** midnight and noon; 2 meters

   b. 6:00 A.M., 6:00 P.M.; 4 meters

   c. midnight to 6:00 A.M., and noon to 6:00 P.M.

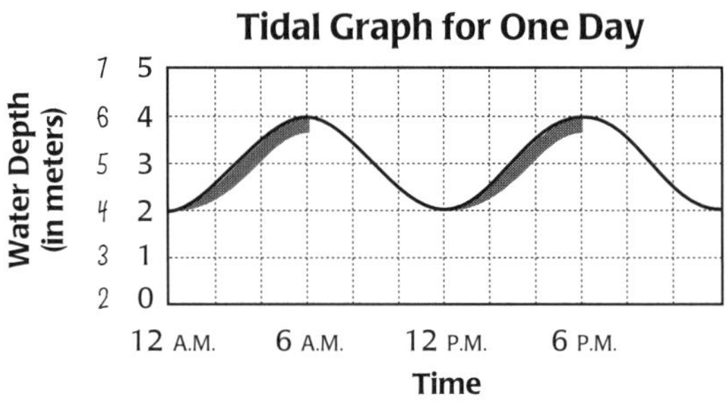

   d. It may enter after 4:00 P.M., and it must leave before 8:00 P.M.

10. The water in this harbor is deeper by 2 meters.

Hints and Comments

Materials Student Activity Sheet 14 (one per student)

Overview Students investigate a new tidal graph, extract information from it, and solve another problem about a ship in the harbor. They then use the same graph to extract information about the tide for a nearby harbor.

Planning You may want students to work individually or in pairs on problems **9** and **10.** Problems **9** and **10** may be used as informal assessments. The Writing Opportunity and/or the Extension on this page may be assigned as homework.

Comments about the Problems

9. Informal Assessment This problem assesses students' ability to extract local, point-specific information from line graphs, as well as to recognize the power of graphs for representing information and solving problems.

a. You may want to check to be sure students understand the scale of the horizontal axis: each square represents a two-hour interval.

c. Be sure students realize that the graph is the wavy line. If they color their answer on the horizontal axis, it is because they were asked to indicate their answer on the graph and did not realize that the graph is the line, not the axis.

10. Informal Assessment This problem assesses students' ability to tell the story depicted by a line graph in terms of the context. Ask students to explain their answers.

Extension You may ask students to make up a problem about the tidal graph of the nearby harbor in problem **10.** They may carry out this activity as homework.

Writing Opportunity You may ask students to describe in their journals some situation in which a tidal graph would be essential. They can use the graph on Student Activity Sheet 14 as an example or illustration. You may choose to have them do this work at home.

Assessment Overview

Students work on four assessment activities that you can use to collect additional information about what each student knows about the major concepts in the unit.

Goals	Assessment Opportunities
• extract local, point-specific information from line graphs	The 1992 Presidential Campaign Trends for Women in the Workforce Barcelona Revisited
• create line graphs from numeric data	Barcelona Revisited Temperature Graphs Revisited
• tell the story depicted by a line graph in terms of the context	The 1992 Presidential Campaign Trends for Women in the Workforce Barcelona Revisited
• interpolate data points to fill in line graphs when appropriate	Barcelona Revisited Temperature Graphs Revisited
• reason about global information such as regularity, maximum/minimum values, slope, and variance as shown in line graphs	The 1992 Presidential Campaign Trends for Women in the Workforce
• create line graphs from verbal descriptions	Temperature Graphs Revisited
• understand limitations of line graphs	Barcelona Revisited
• recognize the importance of scale in creating line graphs	Barcelona Revisited
• recognize the power of graphs for representing information and solving problems	Trends for Women in the Workforce Temperature Graphs Revisited

Pacing

When combined, the four assessment activities will take approximately two 45-minute class sessions. See the Planning Assessment section for further suggestions as to how you might use the assessment activities.

About the Mathematics

The four assessment activities evaluate the major goals of the *Tracking Graphs* unit. Refer to the Goals and Assessment Opportunities section on the previous page for information regarding the specific goals assessed in each assessment activity. Students may use different strategies to solve each problem. Their choice of strategies may indicate their level of comprehension of the problem. Consider how well students' strategies address the problem, as well as how successful students are at applying their strategies in the problem-solving process.

Materials

- Assessments, pages 127–135 of the Teacher Guide (one of each per student)
- Student Activity Sheet 1, page 113 of the Teacher Guide (one of each per student)

Planning Assessment

You may want students to work on these assessments individually so that you can evaluate each student's understanding and abilities. Make sure that you allow enough time for students to complete the assessment activities. Students are free to solve each problem in their own ways. They may choose to use any of the models introduced and developed in this unit to solve problems that do not call for a specific model.

Students can work on the problems in The 1992 Presidential Campaign assessment in pairs. Barcelona Revisited can be used after students have finished Section A. These problems assess almost all the goals of the section. Note that for some students, this assessment may be too difficult at this stage of the unit; they may need more time to develop their abilities and skills. Temperature Graphs Revisited may be used after students have finished Section C.

Scoring

Answers are provided for all assessment problems. The method of scoring the problems depends on the types of questions in each assessment. Most questions require students to explain their reasoning or justify their answers. For these questions, the reasoning used by the students in solving the problems as well as the correctness of the answers should be considered as part of your grading scheme. A holistic scoring scheme can be used to evaluate an entire task. For example, after reviewing a student's work, you may assign a key word such as *emerging, developing, accomplishing,* or *exceeding* to describe his or her mathematical problem solving, reasoning, and communication.

On other tasks, it may be more appropriate to assign point values for each response. Student progress toward the goals of the unit should also be considered. Descriptive statements that include details of a student's solution to an assessment activity can be recorded. These statements would provide insight into a student's progress toward a specific goal of the unit. Descriptive statements can be more informative than recording only a score and can be used to document students' growth in mathematics over time.

THE 1992 PRESIDENTIAL CAMPAIGN Page 1 of 4

Use additional paper as needed.

One of the interesting things about the 1992 presidential campaign was that there were three major candidates, Bill Clinton, George Bush, and Ross Perot. The line graphs on the right show how the popularity of Clinton, Bush, and Perot changed over the seven months before the election. The numbers on the vertical scale are percents of people favoring each candidate.

The Popularity of the Three Candidates

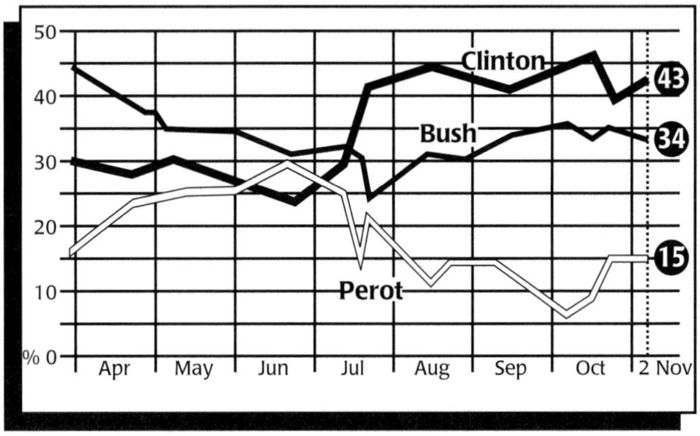

1. a. What were the popularity ratings for each of the three candidates at the beginning of April?

b. What were the popularity ratings for each of the three candidates at the beginning of September?

c. Did each voter polled at the beginning of April and September favor one of the three candidates? Use the information from the graph to support your answer.

Solutions and Samples
of student work

1. **a.** Perot: about 17%

   Clinton: about 30%

   Bush: about 44%

   b. Perot: about 14%

   Clinton: about 43%

   Bush: about 31%

   c. No, the total was 91% in April and 88% in September. In order for all those polled to favor one of these three candidates, the total percent would be 100.

Hints and Comments

Materials The 1992 Presidential Campaign assessment, page 127 of the Teacher Guide (one per student)

Overview Students investigate three line graphs that show how the percent of voters in the 1992 presidential election changed over a period of seven months. Students extract and reason about local information to determine the popularity of the candidates at two points in time.

About the Mathematics Local and global information is extracted from these line graphs. Important features like points of intersection, increase, decrease, maximum, and minimum play a role in these problems and are connected to the context. Actually, this situation is not a continuous function over time, because the data are received from polling. Assuming that there are a very large number of voters, and because the results of the polls are translated into percents, you may consider that all percents between 0% and 100% may occur.

Planning Students may work on problem **1** individually or in pairs.

Comments about the Problems

1. This problem assesses students' ability to extract local, point-specific information from line graphs. You may want to discuss the accuracy that is possible in these answers. The lines of the three graphs are thick and the scale unit is 5%.

   c. Students should know that all voters together represent 100%. You should look for recognition of this fact in their explanation. The remaining voters were either undecided or they favored some other candidate.

THE 1992 PRESIDENTIAL CAMPAIGN

Use additional paper as needed.

Here is a description of Bush's popularity during the period of time shown by the graph on page 1 of this assessment.

Bush started in April with a rating of 44%. His popularity dropped to 25% over halfway through July. He then regained popularity, and after a small dip during the second half of August, he climbed to 36%. In the last weeks of the campaign his popularity went up and down slightly and ended at 34% at the time of the election.

2. Write a description of Clinton's popularity during the same time period.

3. When did Clinton pull ahead of Bush and Perot? How can you tell this from the graph?

4. Describe what happened to the popularity ratings of the three candidates during the month of April.

Solutions and Samples
of student work

2. Descriptions will vary. Sample description:

   Clinton started with 30% in April. During this month he lost some ground, but then regained it. Then he went down to his minimum (23%) halfway through June. His popularity increased after that, and halfway through August, he reached 44%. Then he lost and gained again. His maximum was 47% halfway through October. He lost 6% during the third week of October, but then regained and ended at 43%.

3. Mid-July. After this point, Clinton's graph stays above the other two graphs. Clinton's graph first crosses Perot's and then Bush's. After that time, Clinton is in the lead, which is represented by his graph being above the other two graphs.

4. Bush went down, or decreased in popularity.

   Perot went up, or gained in popularity.

   Clinton lost then regained popularity (over the whole month he lost a little).

Hints and Comments

Materials The 1992 Presidential Campaign assessment, page 128 of the Teacher Guide (one per student)

Overview Students describe one of the graphs in words. They compare the three graphs using relevant points and global information such as increases and decreases.

Planning Students may work individually or in pairs on problems **2–4.**

Comments about the Problems

2–4. These problems assess students' ability to tell the story depicted by a line graph and to reason about global information shown in line graphs.

2. Students should use global and local information from the graph for their story. Look for the essential ideas that should be in any answer: increases and decreases linked to specific times and numbers.

3. To compare the graphs, students have to look more globally. Students may recall how they could tell who was ahead in the running context in Section A, and then use this idea to solve this problem.

4. This problem invites students to look globally at the shape of the graph and translate the "ups and downs" in terms of the context.

THE 1992 PRESIDENTIAL CAMPAIGN Page 3 of 4

Use additional paper as needed.

Use the graph on page 1 of this assessment to answer the following questions.

5. a. During what period of time was Perot more popular than Clinton?

b. Was there ever a period of time when Perot was more popular than Bush? How can you tell this from the graph?

6. a. When did Clinton experience a boost in popularity?

b. By how much did his popularity rise during this period of time?

c. What happened to the popularity ratings of the other two candidates over the same period of time? Explain why this makes sense.

7. There was a point in time during the campaign when the popularity ratings of the three candidates were very close. When was that?

Solutions and Samples
of student work

5. a. during most of the month of June and the beginning of July

b. Perot was never more popular than Bush. The Perot graph is always below that of Bush.

6. a. the second and third week of July

b. by about 13% (from 29% to 42%)

c. Perot lost a lot (going from 25% to 14%). Bush lost a little. Explanations will vary. Some students may reason that about 88% of the voters are still accounted for, so if Clinton gets more voters, they must have come from other candidates, and the other two (or at least one) must be going down.

7. from the first week of June until the middle of July

Hints and Comments

Materials The 1992 Presidential Campaign assessment, page 129 of the Teacher Guide (one per student)

Overview Students compare the three graphs using relevant points and global information such as increases and decreases. They determine the relative popularity of the three candidates at various times during the campaign.

Planning Students may work individually or in pairs on problems **5–7.**

Comments about the Problems

5–7. These problems assess students' ability to extract local and global information from line graphs.

5. Although the wording may differ, this problem is similar to problem **3.** Students have to look for the part of Perot's graph that is higher than the graphs of Clinton or Bush. Then they have to read what time period that part of the graph corresponds to.

6. Students have to look for another important feature of line graphs: the slope or steepness. A larger popularity increase may be shown in the graph by a steeper line. Students' answers will show you how well they understand this concept at this stage. Students will get another opportunity to develop more understanding of the concept of slope and investigate increase and decrease more formally in the grade 7/8 unit *Ups and Downs.*

7. Students find the period during which the three graphs stay close together. This is another example of a problem that requires students to look globally first and then extract local information.

THE 1992 PRESIDENTIAL CAMPAIGN Page 4 of 4

Use additional paper as needed.

During a presidential campaign, a candidate's popularity may increase or decrease. For example, Bill Clinton's popularity decreased during the first part of April and then started to increase again.

The Popularity of the Three Candidates

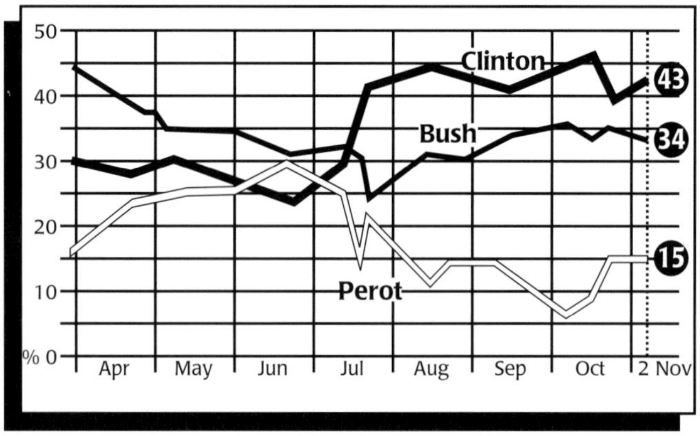

8. a. Indicate when Clinton's popularity changed from increasing to decreasing. How many such times were there? How can you tell this from the graph?

b. Who had the largest increase in popularity from the last week in July to mid-August? How can you tell this from the graph?

Solutions and Samples
of student work

8. a. There are three instances when the graph of Clinton's popularity indicates a change from increasing popularity to decreasing popularity (moving from left to right the direction of the graph switches from upward to downward):

- after the first week of May,
- halfway through August,
- halfway through October.

b. Answers will vary. Some students may say Bush and other students may say Clinton since both candidates experienced increases in popularity (a graph moving upwards to the right) over that time period.

Bush's increase is more than 5% points, whereas Clinton's increase is about 4% points over the same time period.

Hints and Comments

Materials The 1992 Presidential Campaign assessment, page 130 of the Teacher Guide (one per student)

Overview Students examine one candidate's graph to determine when changes in his popularity took place. They compare all the candidates' ratings during a specific period of time.

Planning Students may work on problem **8** individually or in pairs.

Comments about the Problems

8. This problem assesses student's ability to reason about global information shown in line graphs.

a. Students should look at the "corners" of the graph where it changes from "up" to "down." Some students might also identify the corner in mid-July as being a change from increasing to decreasing. It is not, since his popularity continued to increase.

b. In this problem the same concept that applies in problem **6** plays a role: the slope or steepness. In problem **6,** students had to look for the steepest part of one graph; in this problem students have to compare the steepness of three graphs in a given time period.

TRENDS FOR WOMEN IN THE WORKFORCE

Use additional paper as needed.

The line graph on the right shows information regarding the percentage of working women who worked full time from 1970–1990.

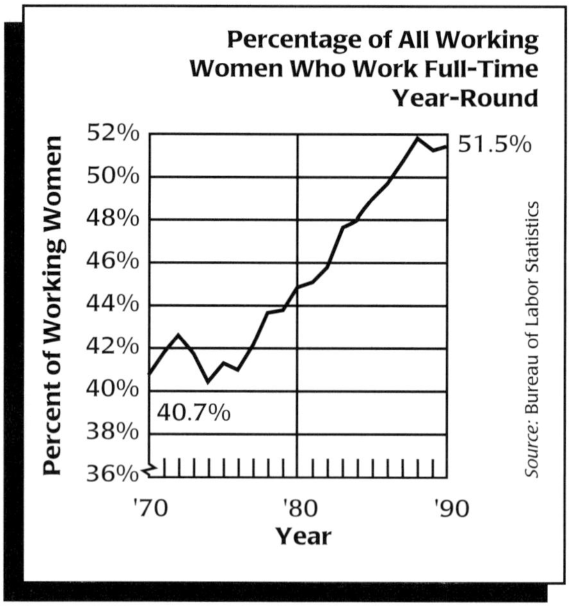

1. Write three questions and their answers based on this graph. One question should be an easy question in your opinion. The other two may be more difficult.

Solutions and Samples
of student work

1. Questions will vary. Sample questions:
   - When is the highest percent reached? [1988]
   - When did the percent go through the 50% barrier? [1986]
   - How much was the percent in the year 1978? [43%]
   - When was the percent 45%? [1980]
   - When did the percent fall after rising? [1972, 1975, and 1988]
   - What is the story of this graph? [In the past fewer than one-half of all working women worked in full-time jobs. But, the trend since the late 70's has been steadily upward. Now, more than one-half of all working women have full-time jobs.]

Hints and Comments

Materials Trends for Women in the Workforce assessment, page 131 of the Teacher Guide (one per student)

Overview Students make up three questions about a given line graph. They also write their answers.

Planning You may want to have students work on problem **1** individually. When students finish their work, you can ask pairs of students to exchange the problems they made up and solve each other's problems. Then they can discuss their solutions. You might want to select a couple of students' questions and compose a new assessment for the whole class.

Comments about the Problems

1. This problem assesses students' ability to extract local and global information from line graphs and to tell the story depicted by a line graph.

   Students' answers will show the level of their understanding of the major concepts of this unit. Encourage students to make up more than just questions about reading data from the graph.

BARCELONA REVISITED

Use additional paper as needed.

At the beginning of the unit, you studied the graph of the men's 100-meter final from the Olympic games in Barcelona. Look again at **Student Activity Sheet l.**

1. Who won the gold (first place), silver (second place), and bronze (third place) medals?

2. How many meters had Dan Mitchell run after 6.48 seconds?

3. In the graph you find the phrases "40 meters" and "4.74 sec."

a. What do these indications mean, and to whom do they apply?

b. How many runners were more than 2 meters behind after 4.74 seconds?

4. Why was a snapshot taken 8.22 seconds after the race began?

Solutions and Samples
of student work

1. Gold: Christie

   Silver: Fredericics

   Bronze: Mitchell

2. a little more than 59 meters

3. **a.** At 4.74 seconds the first runner reached the 40-meter mark. This was B. Surin.

   b. two (the two runners from Nigeria)

4. At that moment the lead runner (Christie) reached the 80-meter mark.

Hints and Comments

Materials Barcelona Revisited assessment, page 132 of the Teacher Guide (one per student); Student Activity Sheet 1 (one per student)

Overview Students extract information from the graph of the 100-meter final from the Olympic games to answer questions about the race.

Planning You may want students to work on problems **1–4** individually.

Comments about the Problems

1–3. These problems assess students' ability to extract local, point-specific information from a line graph.

4. This problem assesses students' ability to tell the story depicted by a line graph.

If students are having difficulty, you may want to point out to them that this problem is a continuation of problem **3a,** except that now the word *snapshot* is used. It might be helpful for the students to remember the idea of the snapshots again when they have to create a graph of one of the runners in the next problem.

BARCELONA REVISITED

Use additional paper as needed.

In this unit you learned how to make line graphs.

5. a. Use the grid on the right to make a line graph for D. Mitchell's 100-meter race.

b. Are there other possible graphs for D. Mitchell's 100-meter race? Why or why not?

6. Imagine that you had a figure that contained the line graphs of all eight runners. Would this figure make the events of the race clearer than the dot graph you have been using? Why or why not?

5. a. Graphs will vary. Sample graph:

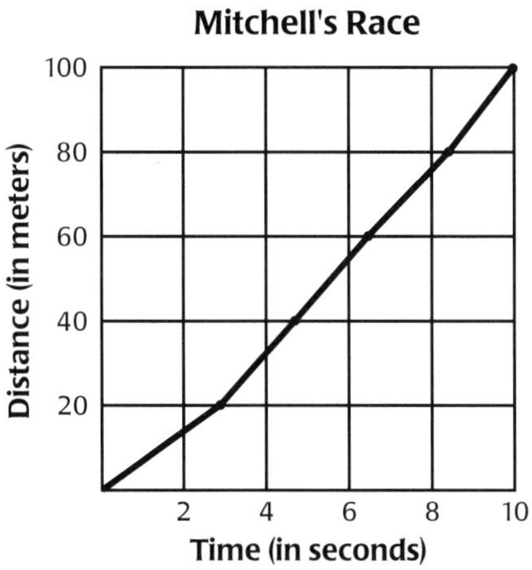

b. Answers will vary. Sample responses:

No. You only have the 6 points. For what happened between these moments, you can only make reasonable guesses.

Yes. Graphs can differ in shape, size, and scale.

6. No. Explanations will vary. Some students may say that, on this scale, the graphs get too close, and you can't see anything clearly. There would be eight graphs and it would be difficult to distinguish one person from another.

Materials Barcelona Revisited assessment, page 133 of the Teacher Guide (one per student)

Overview Students look at one runner's graph and use the information to create a line graph. Then students reflect on two different representations of the race: a diagram with dots, or "snapshots," of the eight runners and a figure showing all eight line graphs.

About the Mathematics Graphs are created with a focus on the axes and the scaling. The scaling of the distances should not be difficult, but scaling the time can be more challenging. You may expect to see students putting the times shown in the diagram at a square distance. All their graphs will become a single straight line.

Comments about the Problems

5. a. This problem assesses students' ability to create line graphs, to interpolate data points to fill in line graphs, and to recognize the importance of scale in creating line graphs.

If students have difficulty, you may remind them of the graph of the race between the two middle schools that they created in Section A: they used six bars and used dots on these bars to indicate the places of the runners. Then they put the bars on a grid and used the dots for drawing the graph. If students are having difficulty with the scaling, you might want to discuss this problem with the whole class.

6. This problem assesses students' understanding of the limitations of line graphs. You may want to ask students to describe one advantage and one disadvantage of the use of the diagram and the graphs.

TEMPERATURE GRAPHS REVISITED Page 1 of 2

Use additional paper as needed.

Consider the following table of temperatures.

Time	Temperature
9:00 A.M.	60°F
12:00 NOON	75°F
3:00 P.M.	84°F
6:00 P.M.	79°F

1. Draw a possible line graph using the data in the table.

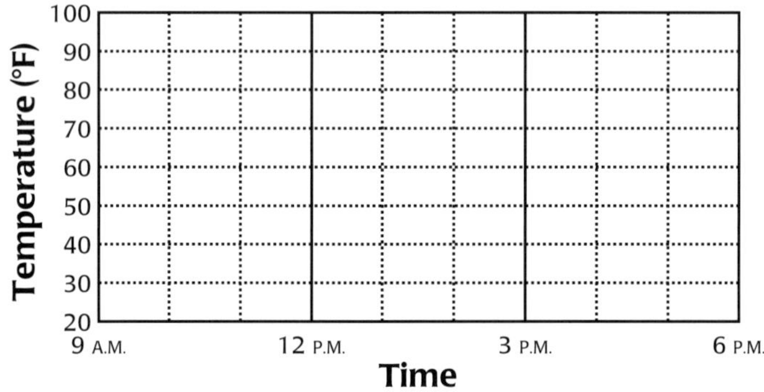

2. Draw another possible line graph, based on the same data, using the additional information that there was a thunderstorm at 1:00 P.M.

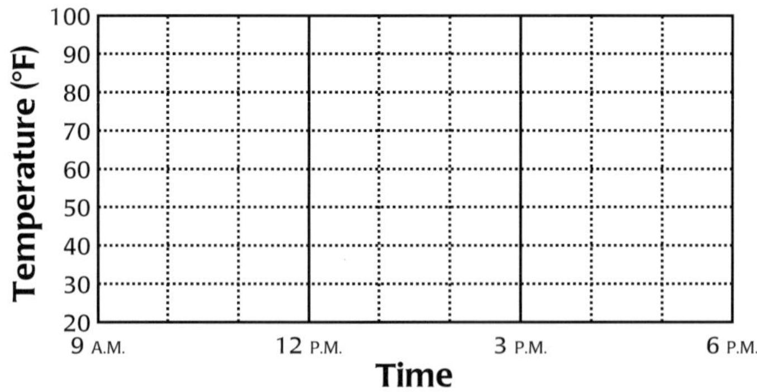

1.

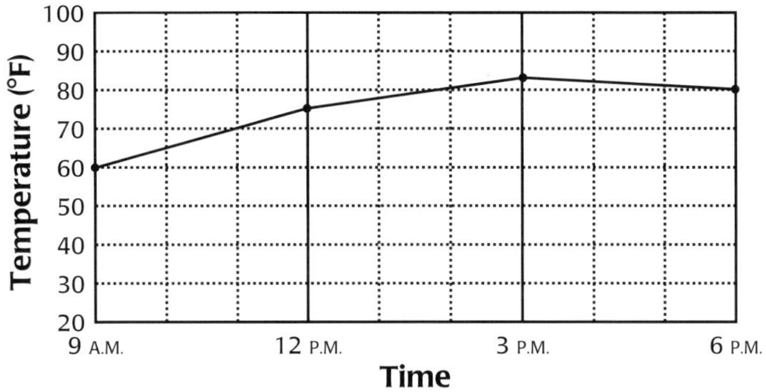

2. Graphs will vary. Sample graph:

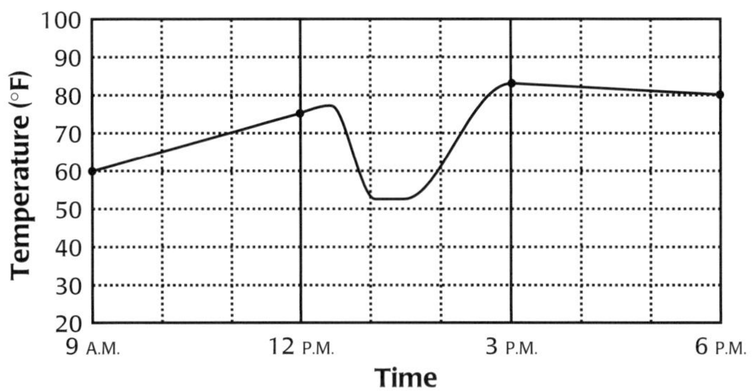

Materials Temperature Graphs Revisited assessment, page 134 of the Teacher Guide (one per student)

Overview Students use a table of temperatures to create a line graph. Then they create another line graph based on additional information.

Planning Students may work on problems **1** and **2** individually.

Comments about the Problems

1. This problem assesses students' ability to create line graphs from numeric data and to interpolate data points to fill in line graphs when appropriate.

All students should be able to create a temperature graph. If students have difficulty, then they should review drawing this type of graph.

TEMPERATURE GRAPHS REVISITED

Use additional paper as needed.

3. Make up a word description of the weather for a week during the summer. The first day has been done for you.

Monday: It was cool in the morning with the temperature at about 55°, but it warmed up by the afternoon to about 80° and stayed nice for the rest of the day.

Tuesday:

Wednesday:

Thursday:

Friday:

Saturday:

Sunday:

4. Make a line graph that represents the temperatures during the week you described.

Week-Long Graph

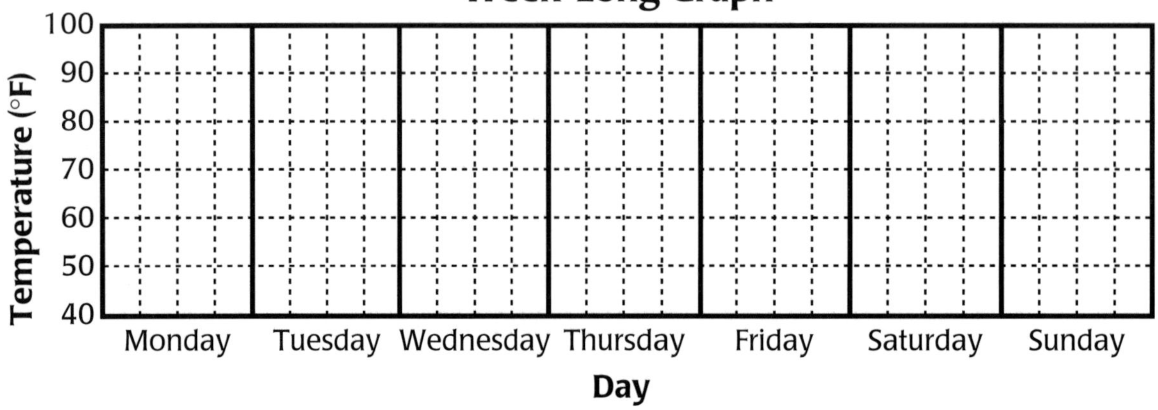

3. Answers will vary. Sample response:

   Tuesday: It was even warmer this day, with the morning temperature about 64°F, and by afternoon it was hot and humid with a high temperature of about 88°F.

   Wednesday: The morning temperature was the same as yesterday's. It was another hot and hazy day with the temperature reaching 91°F.

   Thursday: It was clear early in the morning and 66°F, but it was cloudy by noon, and rain showers kept the temperature at 78°F throughout the afternoon. By evening, the temperature was down to 66°F.

   Friday: A very cool breeze kept the temperature at 60°F at midmorning, and by midafternoon the temperature had only reached 70°F.

   Saturday: A morning low of 52°F was recorded, but it was 72°F by noon and up to 78°F by dinnertime.

   Sunday: It rained in the morning and the temperature was 60°F. It was partly sunny in the afternoon with temperatures around 76°F until early evening.

4. Graphs will vary depending on the weather descriptions that students used for problem **3.** Sample graph:

Week-Long Graph

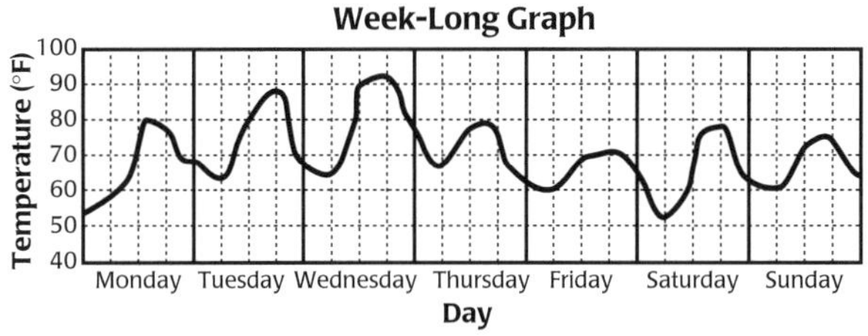

Materials Temperature Graphs Revisited assessment, page 135 of the Teacher Guide (one per student)

Overview Students write descriptions of the weather for six days. They use these descriptions to create a week-long temperature graph.

Comments about the Problems

3. This problem assesses students' ability to recognize the power of graphs for representing information and solving problems.

   Don't worry if the weather described is unlikely to have occurred. The important thing is the agreement of the descriptions and the graph. Answers should include details such as it is usually cooler at night and it warms up during the day. This should be reflected in a wavelike appearance to the graph.

4. This problem assesses students' ability to create line graphs from verbal descriptions.

Extension Ask students to use this week-long graph to create a month-long graph.

Tracking Graphs
Glossary

The Glossary defines all vocabulary words listed on the Section Opener pages. It includes the mathematical terms that may be new to students, as well as words having to do with the contexts introduced in the unit. (Note: The Student Book has no glossary in order to allow students to construct their own definitions, based on their personal experiences with the unit activities.)

The definitions below are specific to the use of the terms in this unit. The page numbers given are from this Teacher Guide.

graph mask (p. 16) A (paper) construction made by students. A line graph may be inserted into the graph mask and "pulled" through it. The gap in the mask allows students to focus on one point on the graph at a time, and by moving the graph through the mask, students can see how the height of the graph changes over time.

line graph (p. 22) a special kind of graph that represents information occurring over time

Blackline Masters

Letter to the Family

Dear Family,

Your child will soon begin the *Mathematics in Context* unit *Tracking Graphs.* Below is a letter to your child that opens the unit, describing the unit and its goals.

In this unit, students use graphs to make sense of the world around them. They will study data from track races, temperature data, and the schedules of oceanic tides and then use graphs to better understand these situations.

You can assist your child through a variety of at-home activities. You might share the temperature graph that is published monthly in most newspapers or ask them about other graphs such as those found in the business section.

You might talk about how you use graphs in your work. Perhaps graphs also help you make decisions at home—some electric companies provide graphs on billing statements.

We hope both you and your child will enjoy the unit and find it a worthwhile experience.

Sincerely,

The Mathematics in Context Development Team

Dear Student,

Did you ever wonder about the tides of the ocean? Did you ever think about how much the temperature changes in one year? Did you ever investigate the way people run races? In this unit, *Tracking Graphs*, you will use graphs to investigate tides, temperatures, and races. Graphs will help you better understand many interesting things.

In this unit, you will be given information about different situations, and you will use graphs to gain further understanding.

Once you know how to make and interpret graphs, you will find it easier to draw conclusions and explain various situations. This will make you a very powerful mathematical thinker.

Sincerely,

The Mathematics in Context Development Team

Name____________________________________

Student Activity Sheet 1

Use with *Tracking Graphs*, pages 1, 2, 3, and 9.

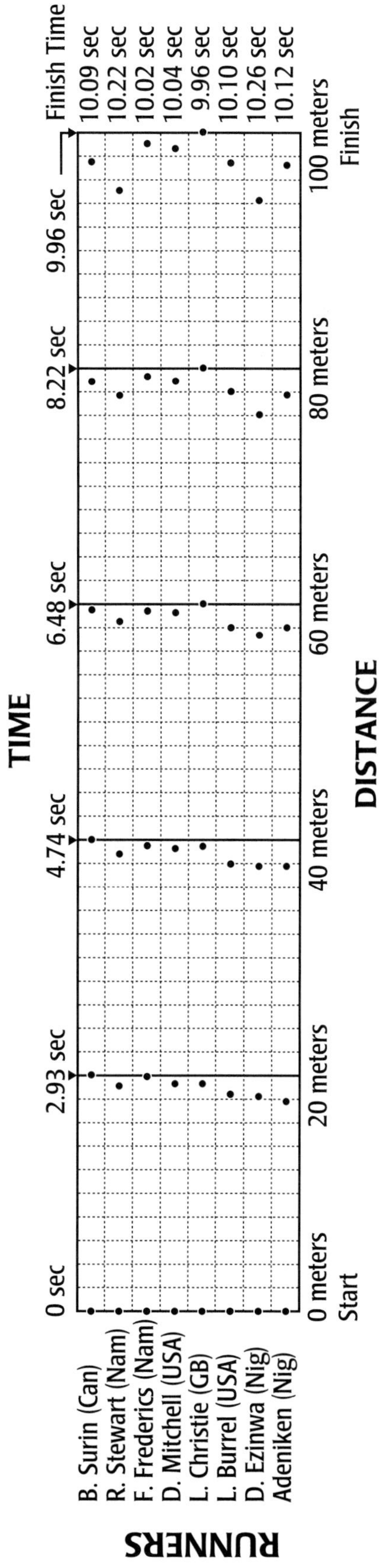

Student Activity Sheet 2

Name ____________________________

Use with *Tracking Graphs*, pages 5 and 6.

Figure 1: 3 minutes

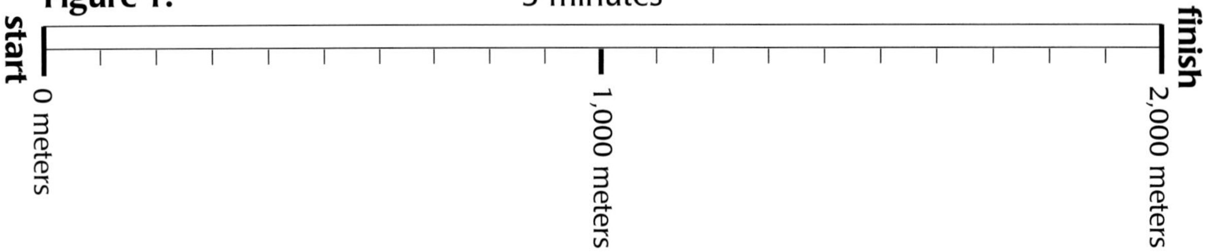

Figure 2: 5 minutes

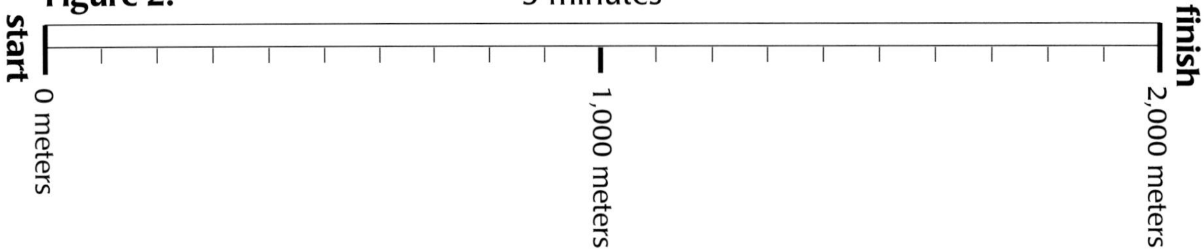

Figure 3: 7.5 minutes

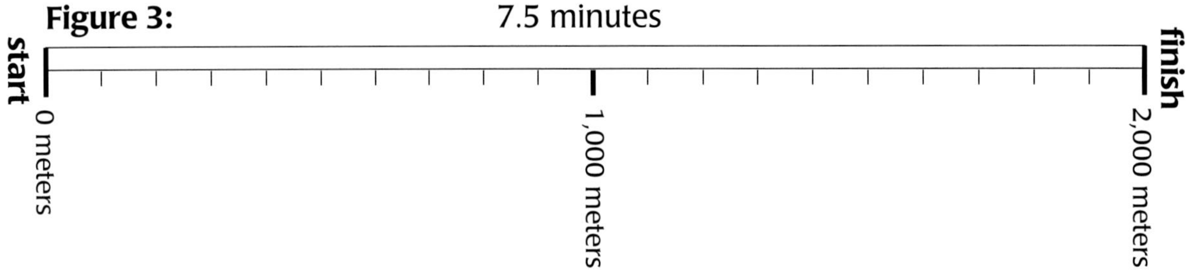

Figure 4: 9 minutes

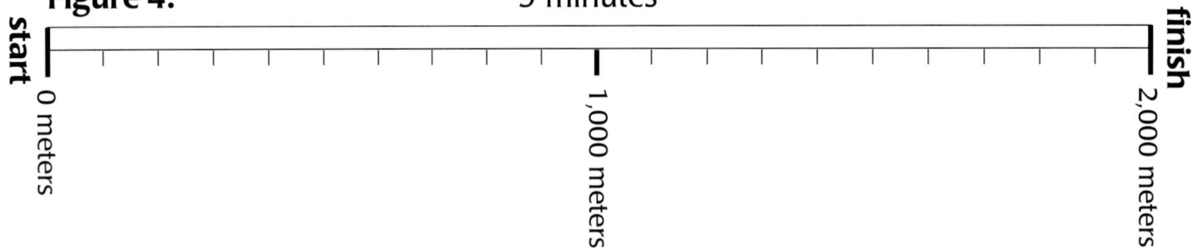

Figure 5: 10 minutes, the finish

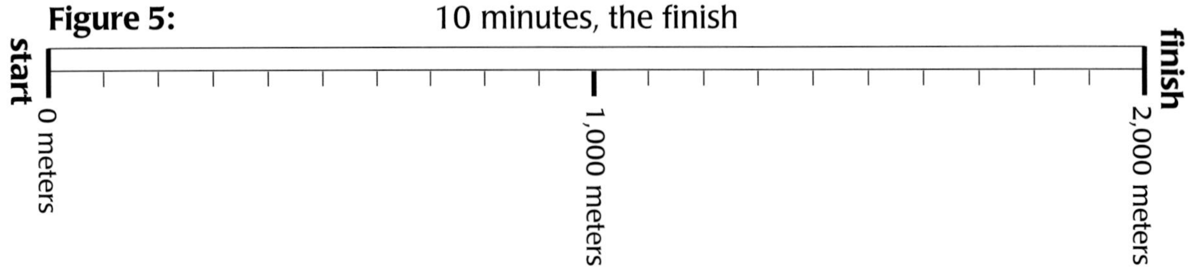

Name________________________________

Student Activity Sheet 3

Use with *Tracking Graphs*, pages 6 and 7.

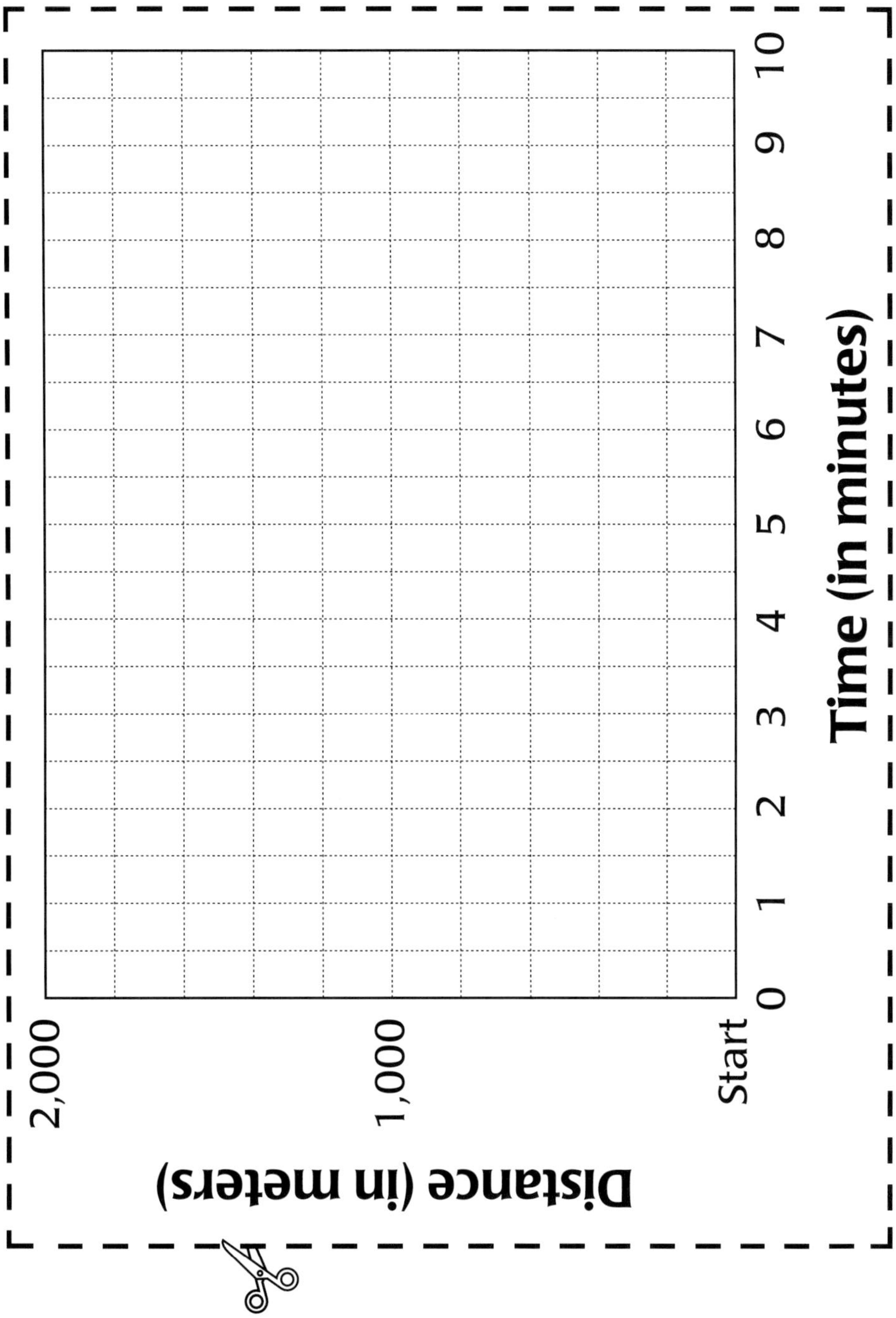

Student Activity Sheet 4

Name ______________________________

Use with *Tracking Graphs*, page 6.

❶ Cut out rectangle on dashed line

❷ Fold on center line

❸ Cut out shaded center strip

❹ Unfold sheet

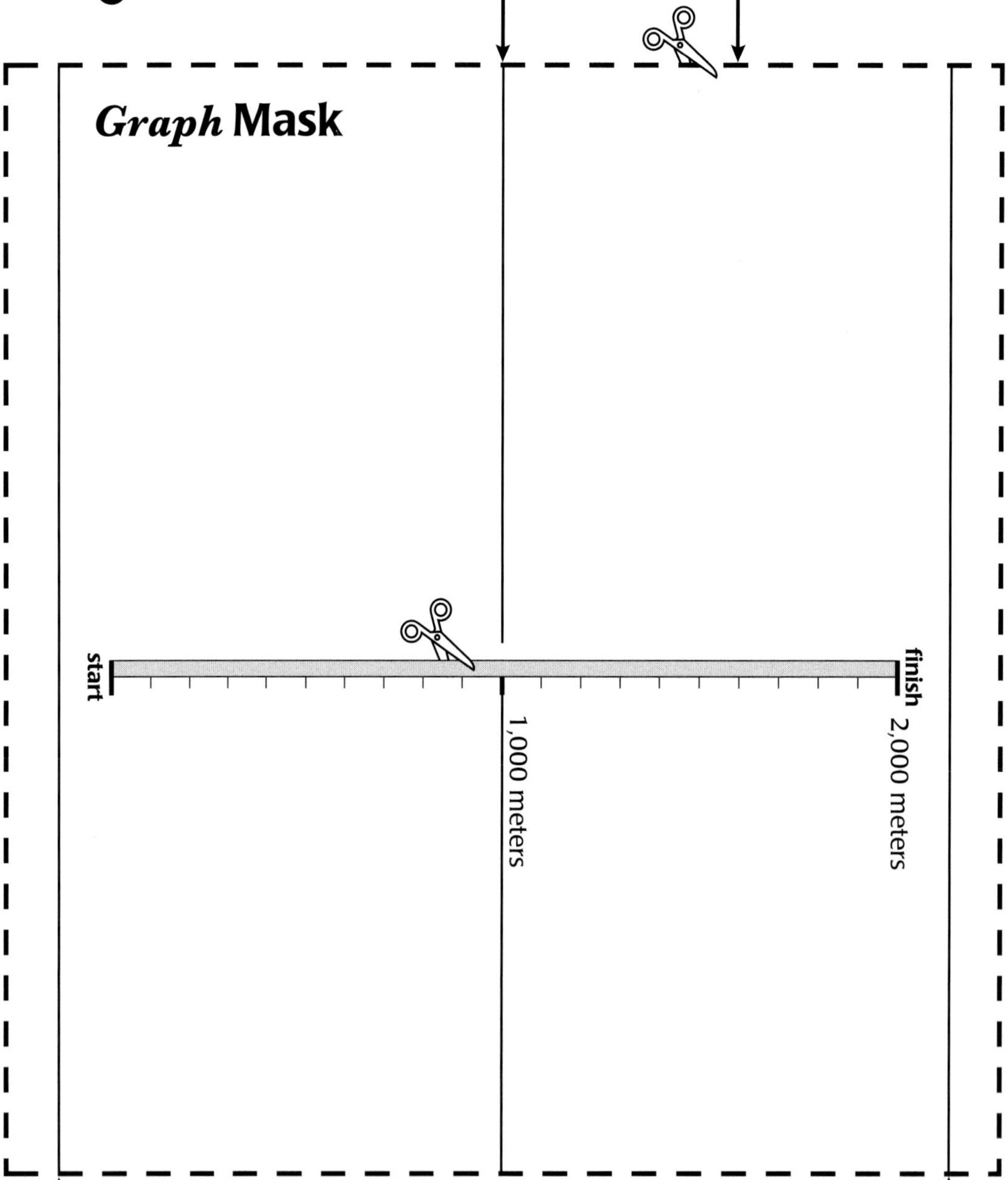

❺ Fold down sides

Name____________________________________

Student Activity Sheet 5

Use with *Tracking Graphs,* page 12.

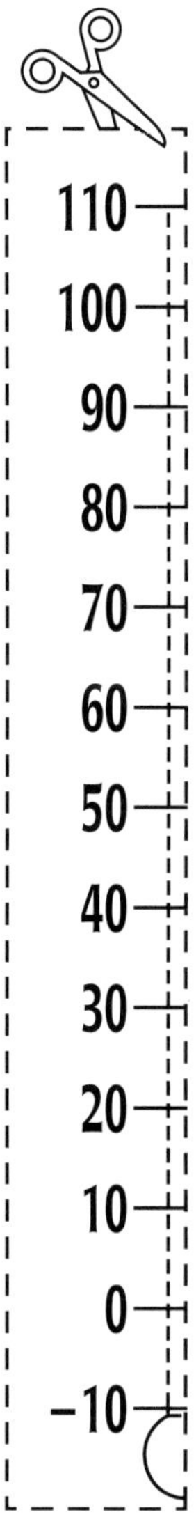

Student Activity Sheet 6

Name ______________________________

Use with *Tracking Graphs*, pages 12, 13, and 14.

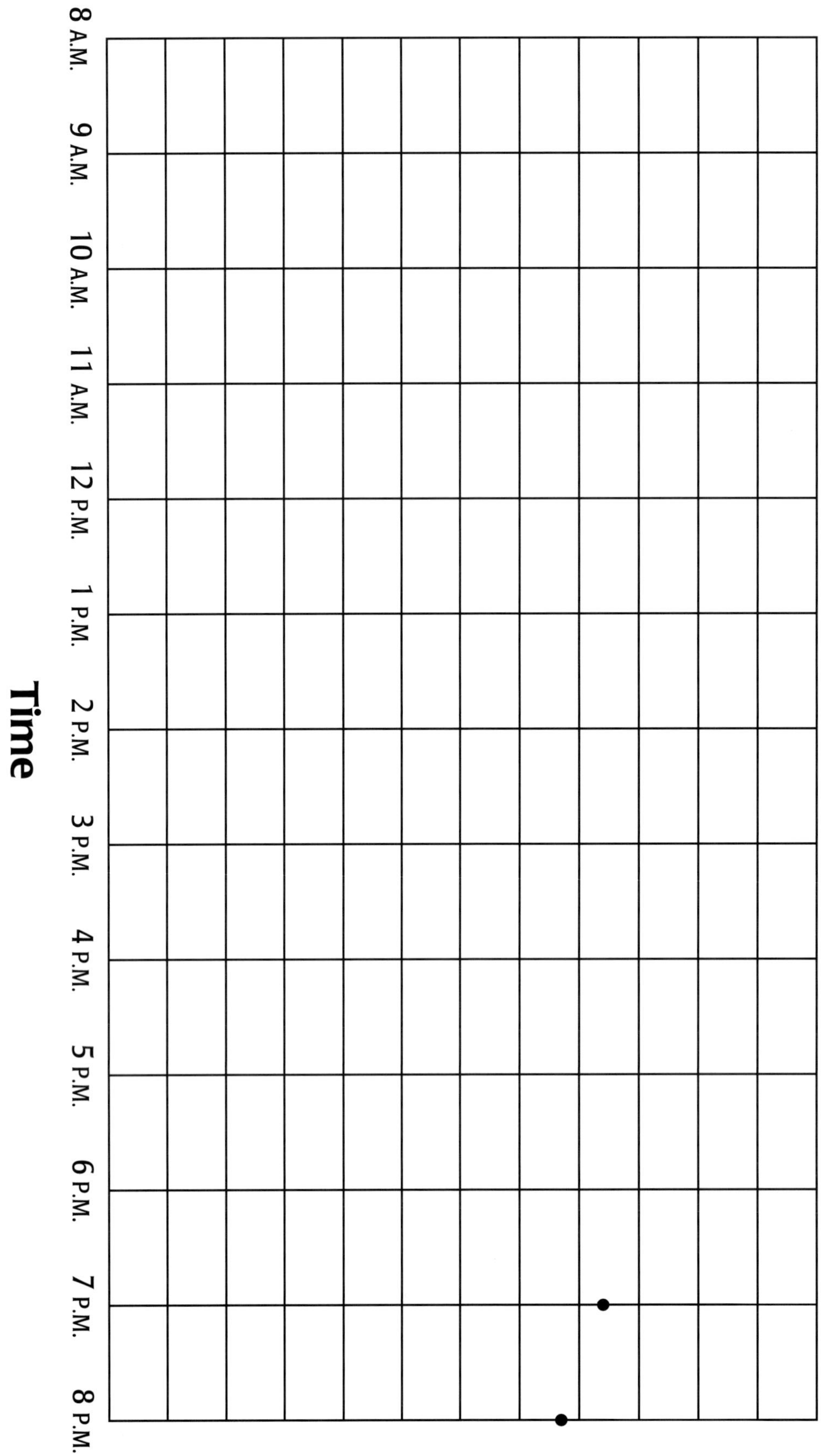

Name________________________________

Student Activity Sheet 7

Use with *Tracking Graphs,* page 15.

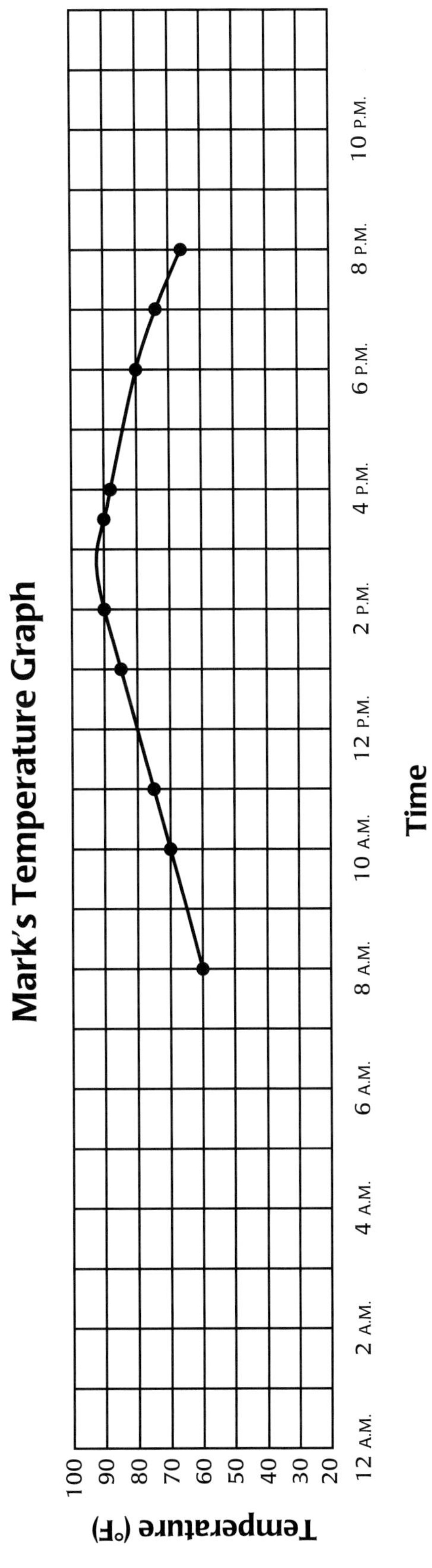

Student Activity Sheet 8

Name ____________________

Use with *Tracking Graphs,* page 16.

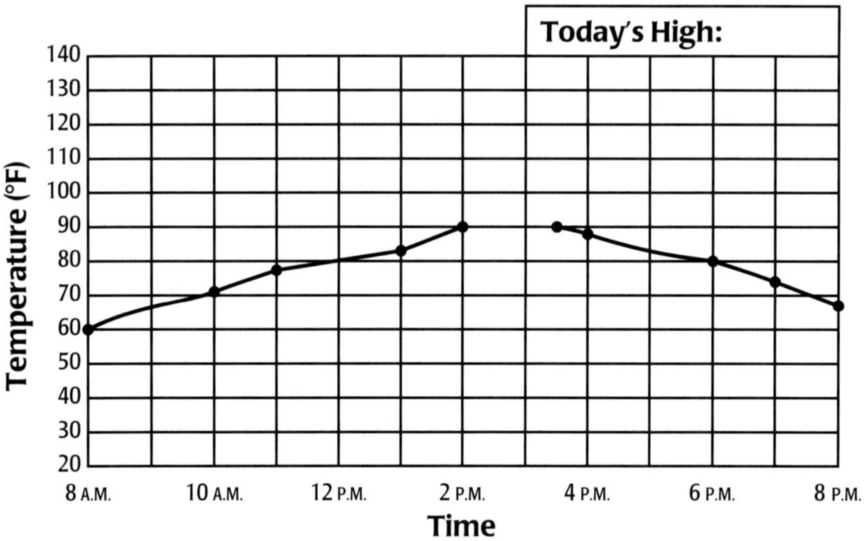

Possibility a:

Between 2:00 P.M. and 3:30 P.M., the temperature stayed at 90°F.

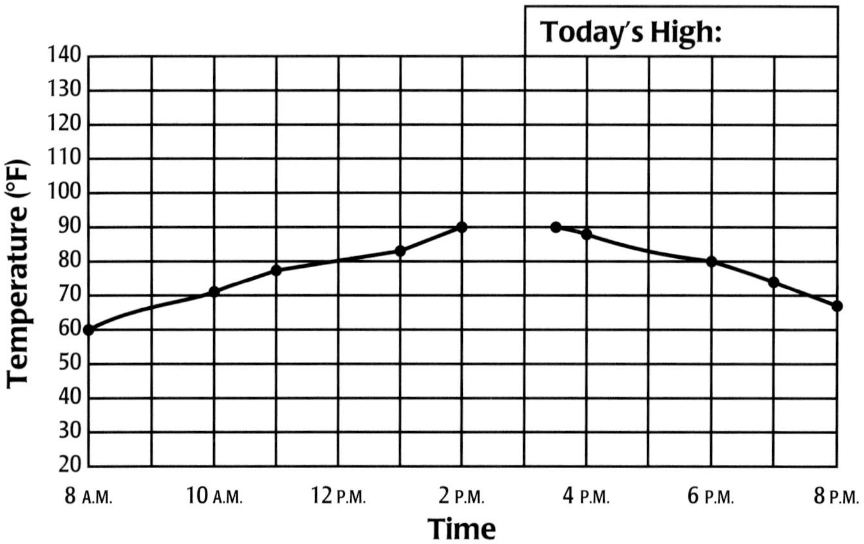

Possibility b:

Between 2:00 P.M. and 3:30 P.M., it got even hotter than 90°F.

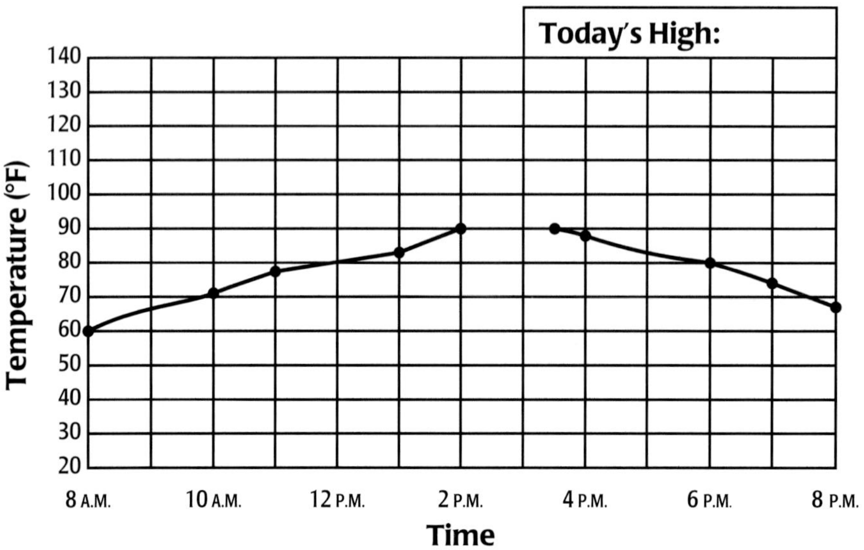

Possibility c:

After 2:00 P.M. there was a severe thunderstorm, which caused the temperature to drop. When the storm was over, the temperature rose again. At 3:30 P.M., it was again 90°F.

Name____________________________________

Student Activity Sheet 9

Use with *Tracking Graphs*, page 20.

Time	°F	°C
8 A.M.	60	
10 A.M.	67	
12 P.M.	75	
2 P.M.	50	
4 P.M.	58	
6 P.M.	57	

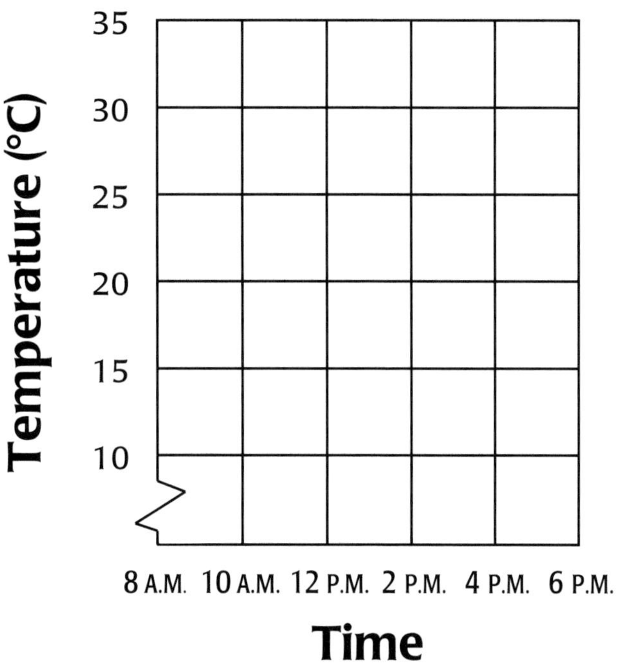

Student Activity Sheet 10

Name ______________________________

Use with *Tracking Graphs*, pages 21, 22, and 23.

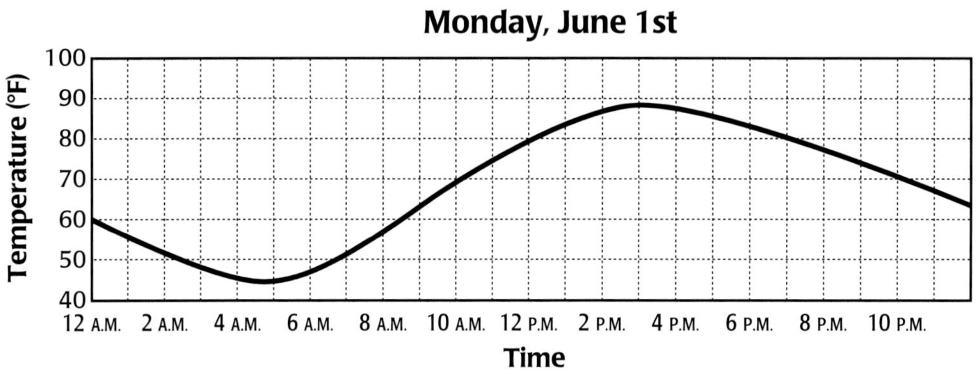

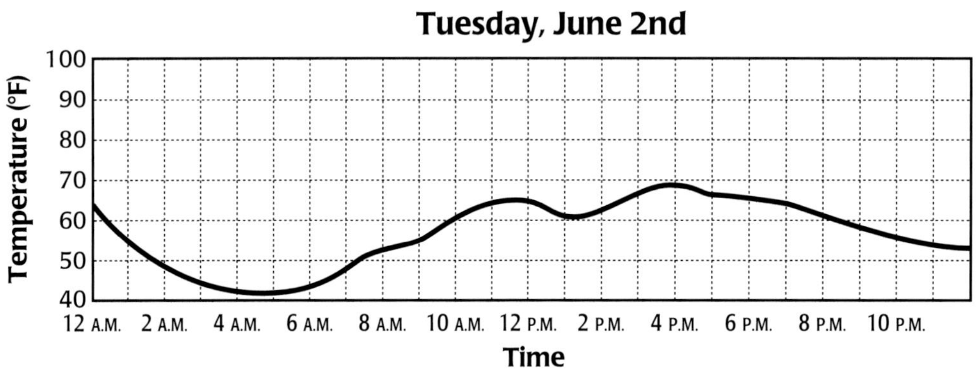

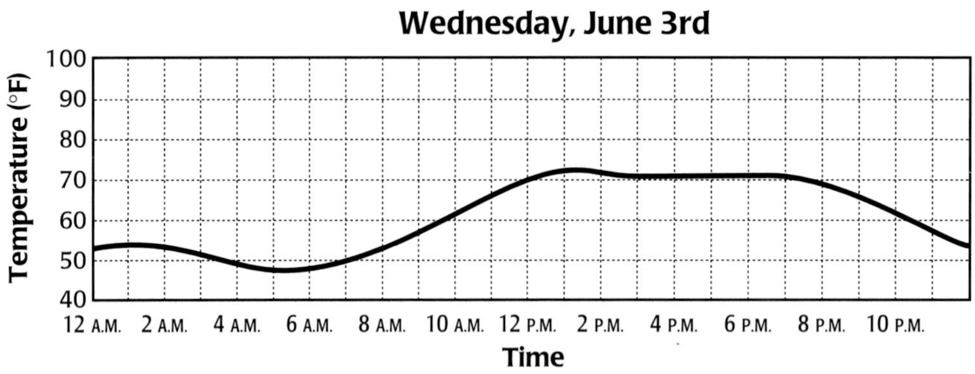

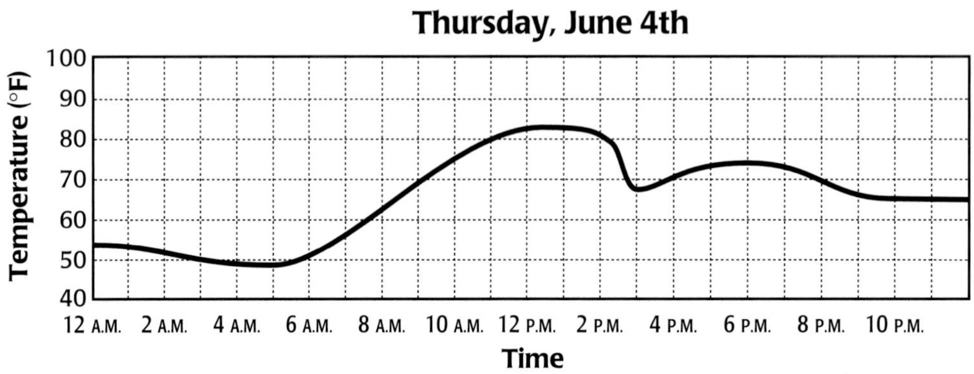

Name____________________________________

Student Activity Sheet 11

Use with *Tracking Graphs,* pages 21, 22, and 23.

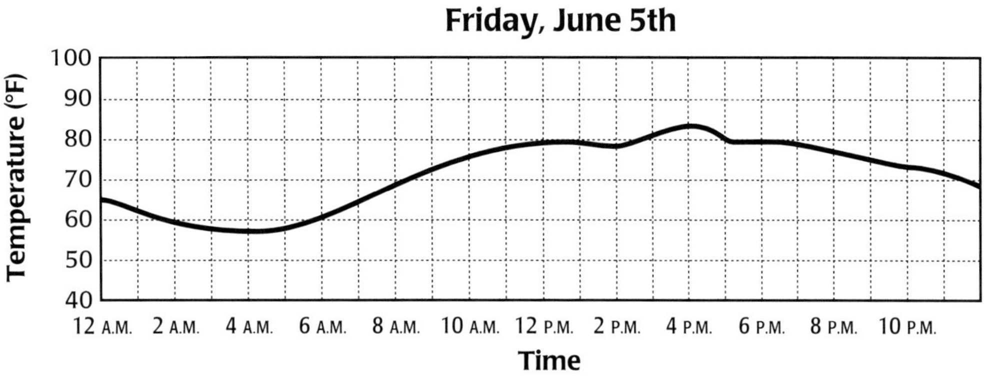

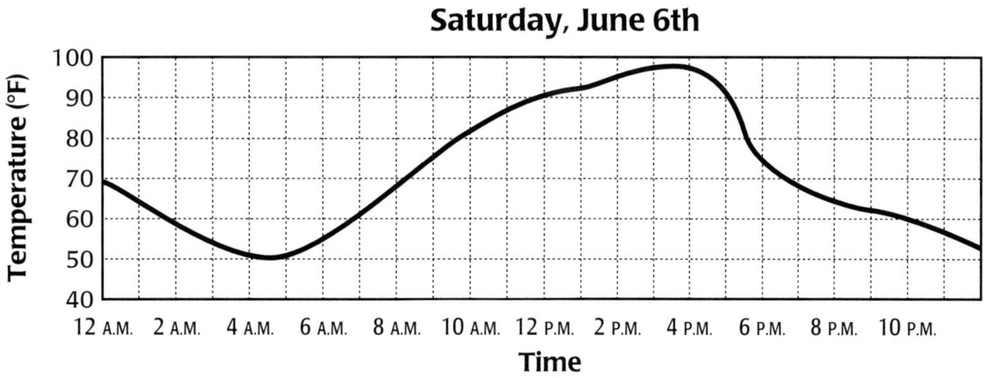

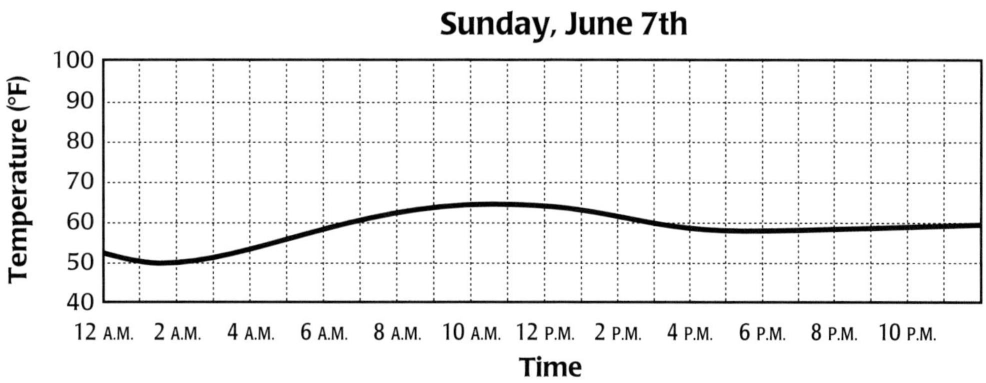

Student Activity Sheet 12

Name ______________________________

Use with *Tracking Graphs*, pages 24 and 25.

Week-Long Graph

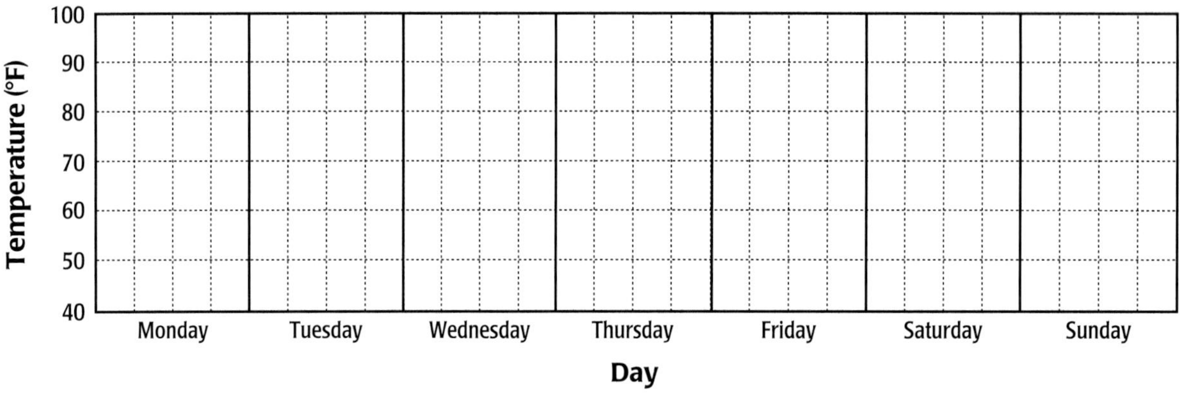

Month-Long Graph

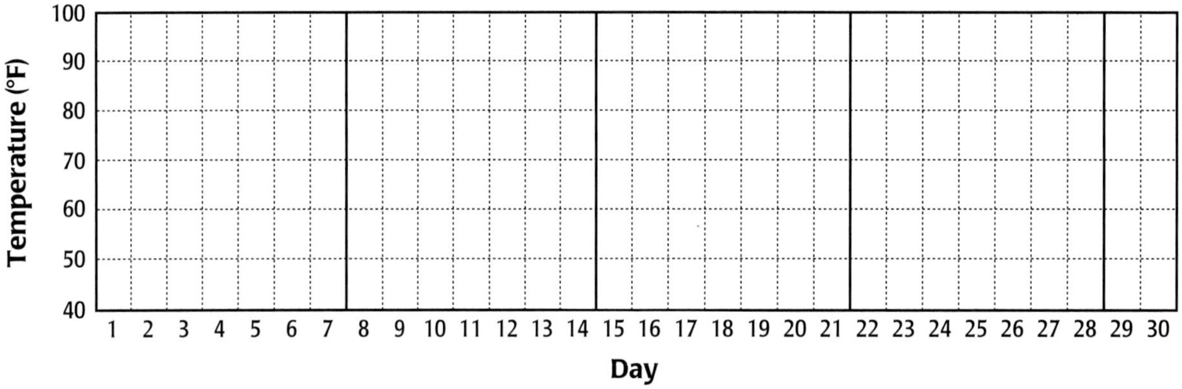

Year-Long Graph

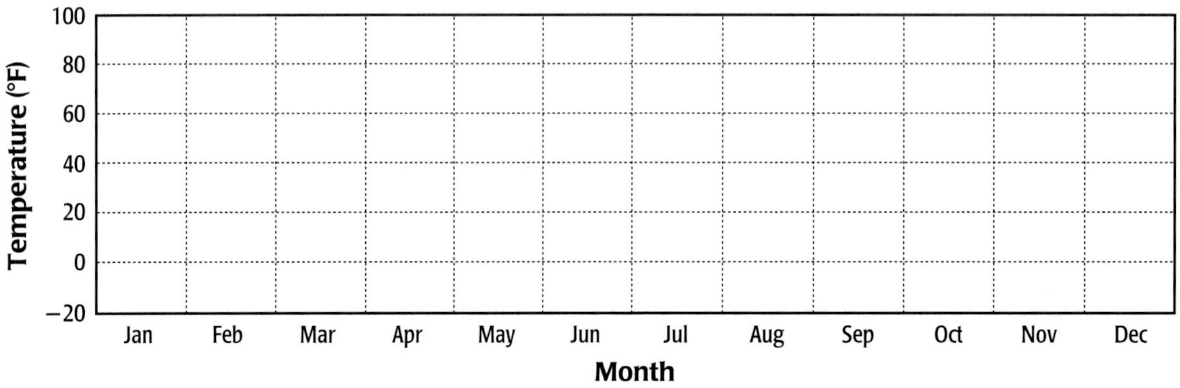

Name________________________________

Student Activity Sheet 13

Use with *Tracking Graphs,* pages 33, 34, 36, and 37.

Tidal Graph for One Day

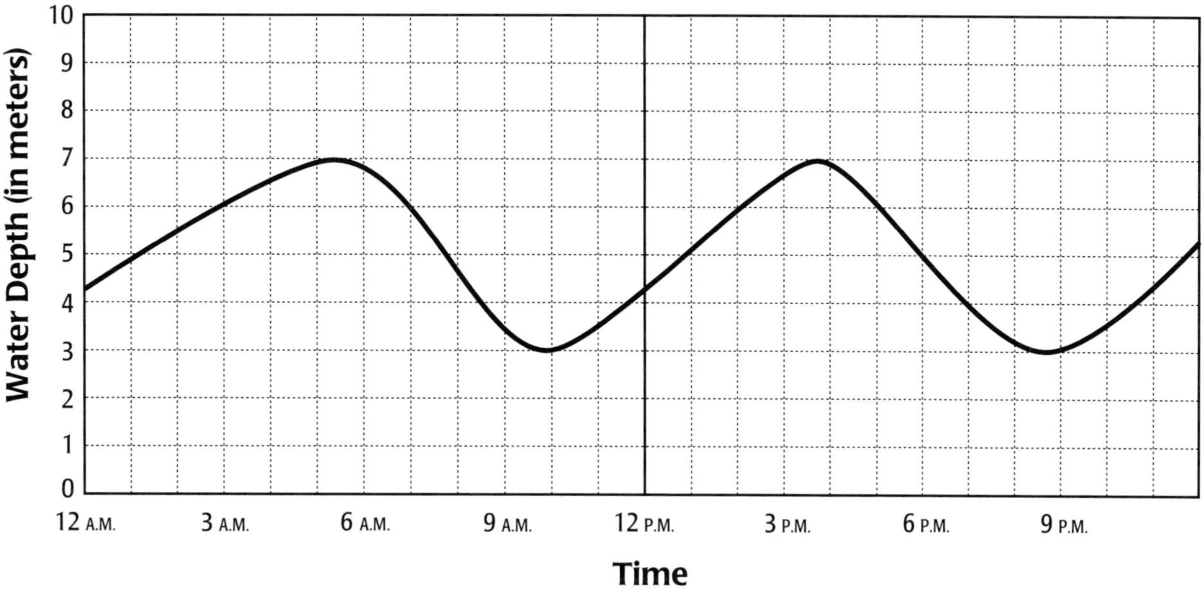

Plant Harbor—Tidal Graph for One Day

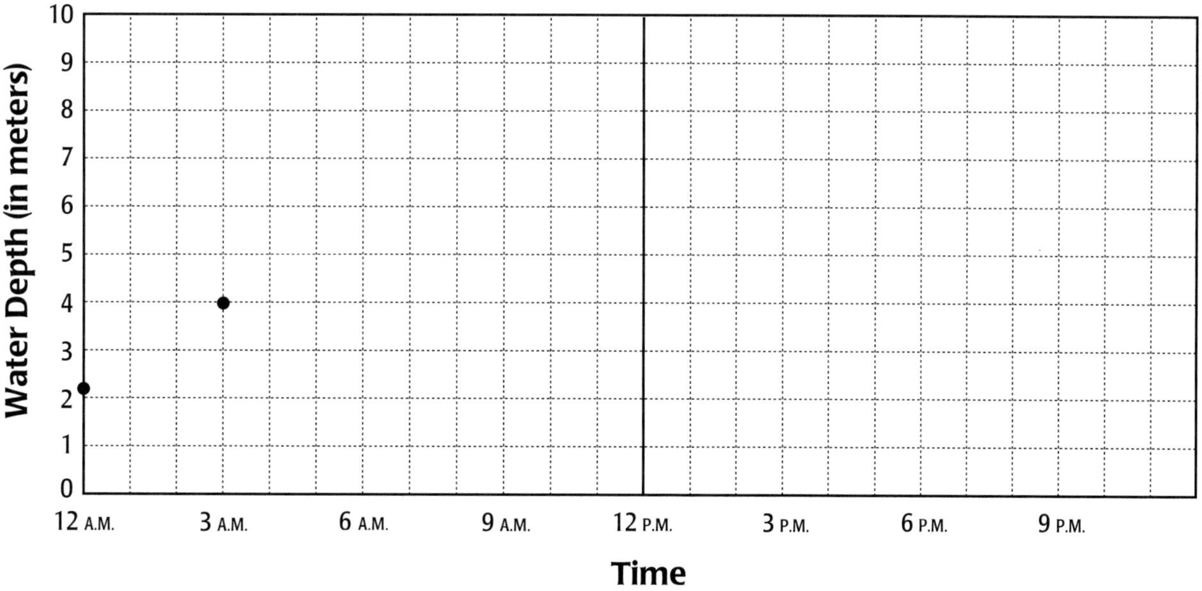

Student Activity Sheet 14

Name ______________________

Use with *Tracking Graphs*, page 38.

Tidal Graph for One Day

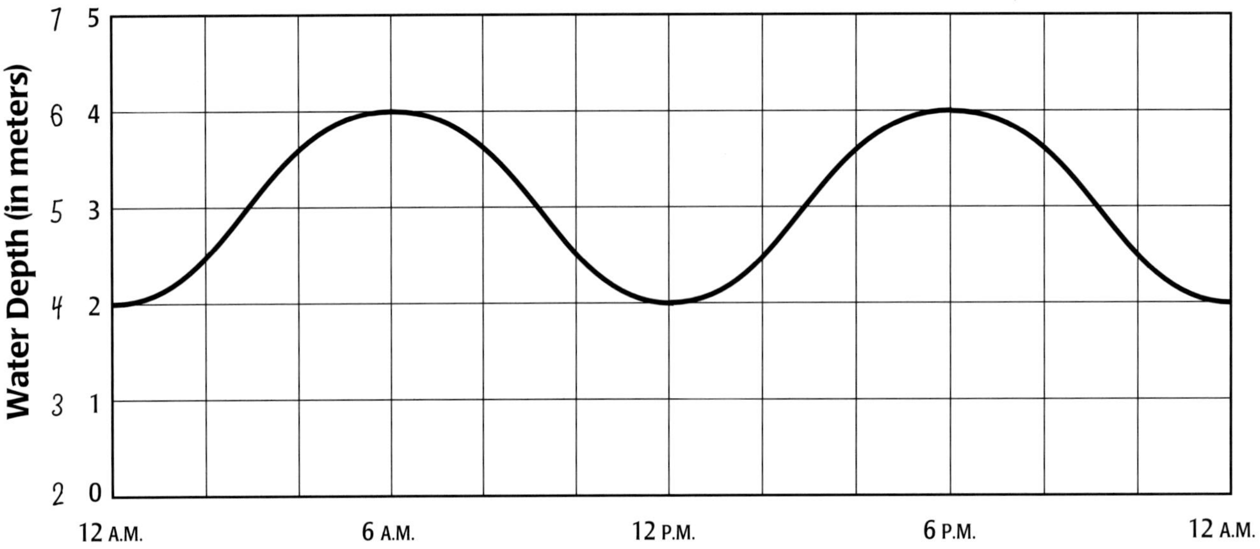

Name________________________ Date________________________

THE 1992 PRESIDENTIAL CAMPAIGN

Use additional paper as needed.

One of the interesting things about the 1992 presidential campaign was that there were three major candidates, Bill Clinton, George Bush, and Ross Perot. The line graphs on the right show how the popularity of Clinton, Bush, and Perot changed over the seven months before the election. The numbers on the vertical scale are percents of people favoring each candidate.

The Popularity of the Three Candidates

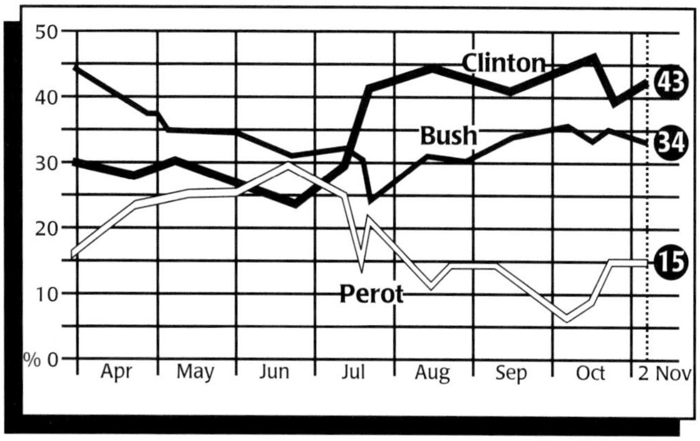

1. a. What were the popularity ratings for each of the three candidates at the beginning of April?

b. What were the popularity ratings for each of the three candidates at the beginning of September?

c. Did each voter polled at the beginning of April and September favor one of the three candidates? Use the information from the graph to support your answer.

Name________________________________ Date____________________

THE 1992 PRESIDENTIAL CAMPAIGN

Use additional paper as needed.

Here is a description of Bush's popularity during the period of time shown by the graph on page 1 of this assessment.

Bush started in April with a rating of 44%. His popularity dropped to 25% over halfway through July. He then regained popularity, and after a small dip during the second half of August, he climbed to 36%. In the last weeks of the campaign his popularity went up and down slightly and ended at 34% at the time of the election.

2. Write a description of Clinton's popularity during the same time period.

3. When did Clinton pull ahead of Bush and Perot? How can you tell this from the graph?

4. Describe what happened to the popularity ratings of the three candidates during the month of April.

Name ______________________ Date ______________________

Use additional paper as needed.

Use the graph on page 1 of this assessment to answer the following questions.

5. a. During what period of time was Perot more popular than Clinton?

b. Was there ever a period of time when Perot was more popular than Bush? How can you tell this from the graph?

6. a. When did Clinton experience a boost in popularity?

b. By how much did his popularity rise during this period of time?

c. What happened to the popularity ratings of the other two candidates over the same period of time? Explain why this makes sense.

7. There was a point in time during the campaign when the popularity ratings of the three candidates were very close. When was that?

Use additional paper as needed.

During a presidential campaign, a candidate's popularity may increase or decrease. For example, Bill Clinton's popularity decreased during the first part of April and then started to increase again.

The Popularity of the Three Candidates

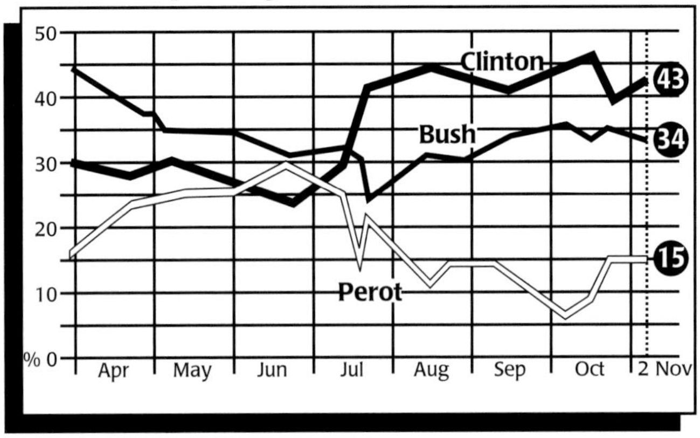

8. a. Indicate when Clinton's popularity changed from increasing to decreasing. How many such times were there? How can you tell this from the graph?

b. Who had the largest increase in popularity from the last week in July to mid-August? How can you tell this from the graph?

Name ______________________ Date ______________________

TRENDS FOR WOMEN IN THE WORKFORCE

Use additional paper as needed.

The line graph on the right shows information regarding the percentage of working women who worked full time from 1970–1990.

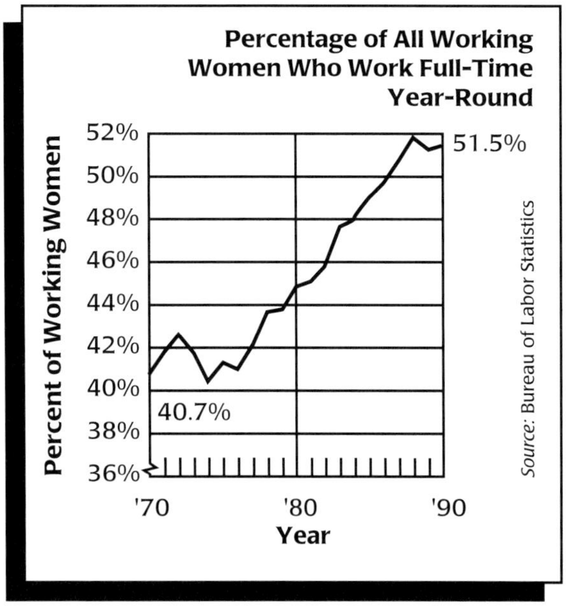

1. Write three questions and their answers based on this graph. One question should be an easy question in your opinion. The other two may be more difficult.

Name ______________________________ Date ______________________

BARCELONA REVISITED

Use additional paper as needed.

At the beginning of the unit, you studied the graph of the men's 100-meter final from the Olympic games in Barcelona. Look again at **Student Activity Sheet 1.**

1. Who won the gold (first place), silver (second place), and bronze (third place) medals?

2. How many meters had Dan Mitchell run after 6.48 seconds?

3. In the graph you find the phrases "40 meters" and "4.74 sec."

a. What do these indications mean, and to whom do they apply?

b. How many runners were more than 2 meters behind after 4.74 seconds?

4. Why was a snapshot taken 8.22 seconds after the race began?

Name ________________________ Date ________________

Use additional paper as needed.

In this unit you learned how to make line graphs.

5. a. Use the grid on the right to make a line graph for D. Mitchell's 100-meter race.

b. Are there other possible graphs for D. Mitchell's 100-meter race? Why or why not?

6. Imagine that you had a figure that contained the line graphs of all eight runners. Would this figure make the events of the race clearer than the dot graph you have been using? Why or why not?

Name ______________________ Date ______________________

TEMPERATURE GRAPHS REVISITED

Use additional paper as needed.

Consider the following table of temperatures.

Time	Temperature
9:00 A.M.	60°F
12:00 NOON	75°F
3:00 P.M.	84°F
6:00 P.M.	79°F

1. Draw a possible line graph using the data in this table.

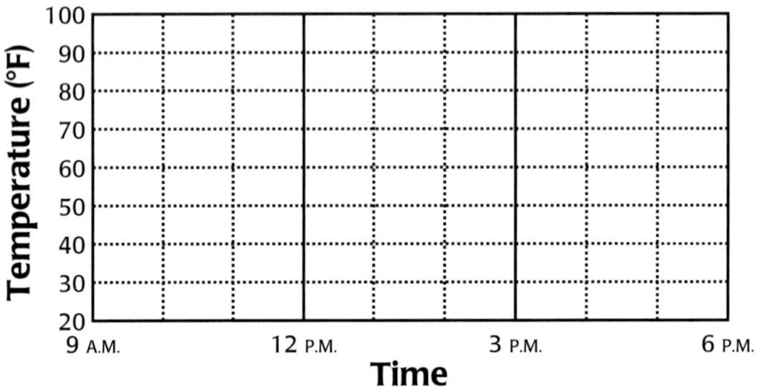

2. Draw another possible line graph, based on the same data, using the additional information that there was a thunderstorm at 1:00 P.M.

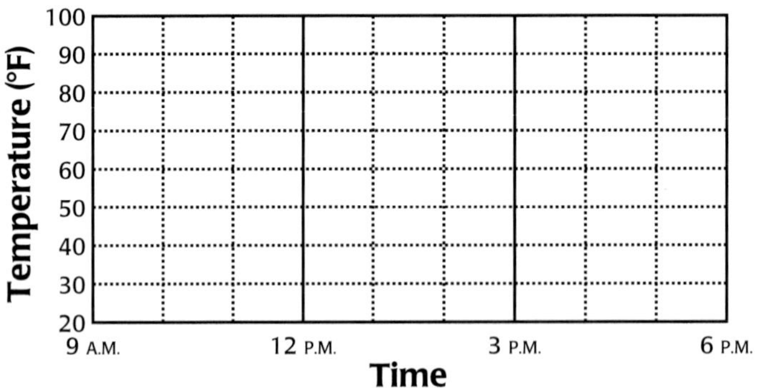

Name ______________________ Date ______________________

TEMPERATURE GRAPHS REVISITED

Use additional paper as needed.

3. Make up a word description of the weather for a week during the summer. The first day has been done for you.

Monday: It was cool in the morning with the temperature at about 55°, but it warmed up by the afternoon to about 80° and stayed nice for the rest of the day.

Tuesday:

Wednesday:

Thursday:

Friday:

Saturday:

Sunday:

4. Make a line graph that represents the temperatures during the week you described.

Week-Long Graph

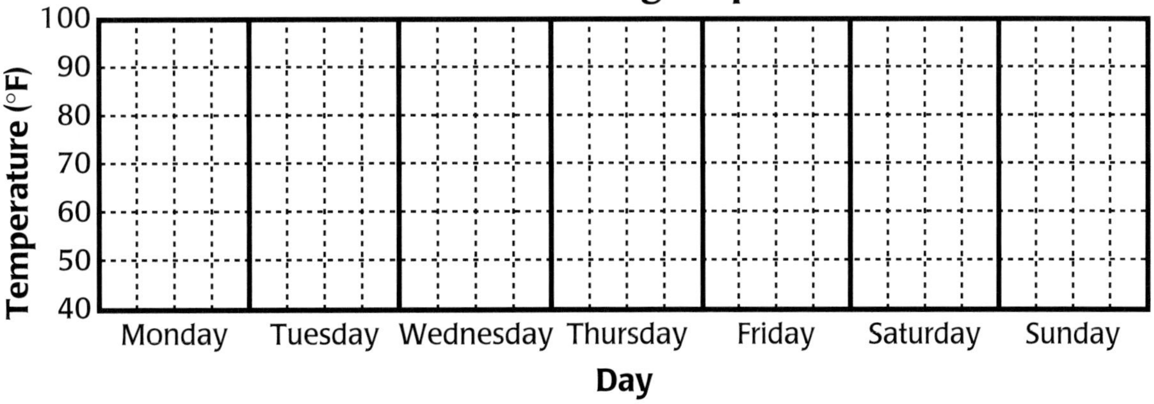

Section A. Graphing a Race

1. **a.** Cyclist B won the race in about 2 hours and 50 minutes. The graph of cyclist B reaches the 50-kilometer distance line first.

   b. Cyclist A had the fastest start. Students may note that the graph of cyclist A is above the other two graphs, or that after half an hour, cyclist A had cycled the greatest distance.

   c. Answers will vary. Some students may say that cyclist B had to stop because of mechanical difficulties.

   d. At 25 kilometers, the riders hit the hill; every graph levels off dramatically at this point.

   e. Estimates will vary. Some students may say that if the graph continued in this way, then cyclist C probably finished in $3\frac{1}{2}$ to 4 hours.

2. Articles will vary.

   Students' answers should include the following:

   - information about the total distance of the race,
   - the name and time of the winner,
   - the information that C is behind all the time,
   - the information that B passes A after 1 hour and 40 minutes, after being very close behind A for 30 minutes, and
   - the information that after 2 hours, B had a short stop.

Section B. Filling in Temperatures

1. Answers will vary. Sample responses:

Day 1 On day 1 the height of the river was about 513 centimeters until it started to rain in the afternoon. At 4 P.M. the river began to rise. At 12 A.M. the height was already 575 centimeters.

Day 2 During the night, it stopped raining, but the water level didn't drop very much. It stayed at a height of about 575 centimeters during that day.

Day 3 At 4 A.M. it started raining again, and the water level went from about 575 to 620 centimeters. It stayed at this level from about 10 A.M. until 6 P.M. Then it went up again, possibly due to the water that came from a higher mountain area.

Day 4 During the night the water level stayed at about 630 cm. Then around 8 A.M., the water level dropped visibly. At 8 P.M. the water level seemed to stabilize at 575 centimeters.

2. a.

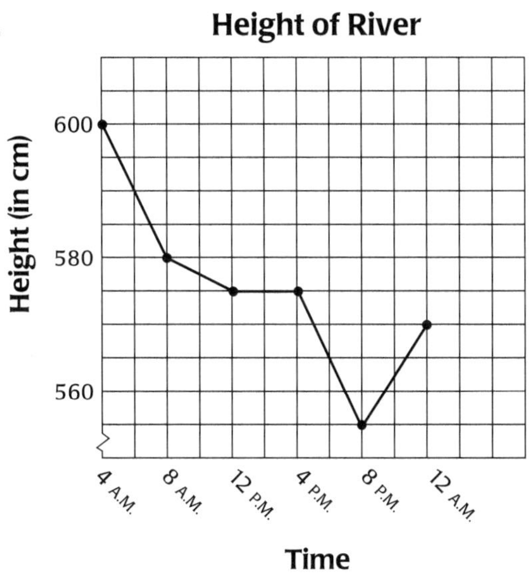

b. Graphs will vary, depending on how the river rises and falls between 12:00 P.M. and 4 P.M.

Sample graph:

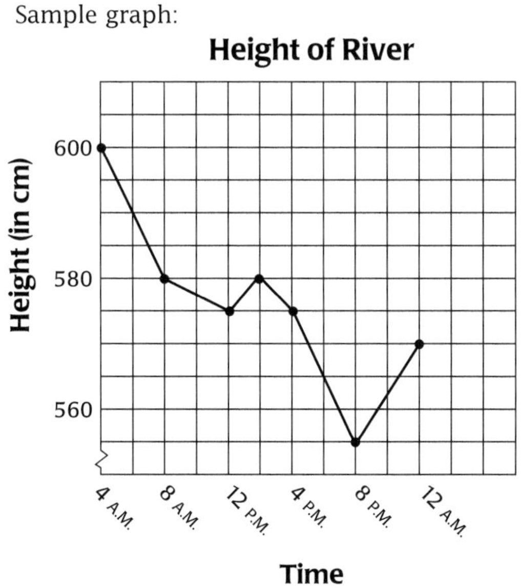

Section C. Graphs over a Longer Time Period

1. a. a little less than 20 pounds

b. During the month of April she gained about 4 pounds. In September, she also gained about 4 pounds.

2.

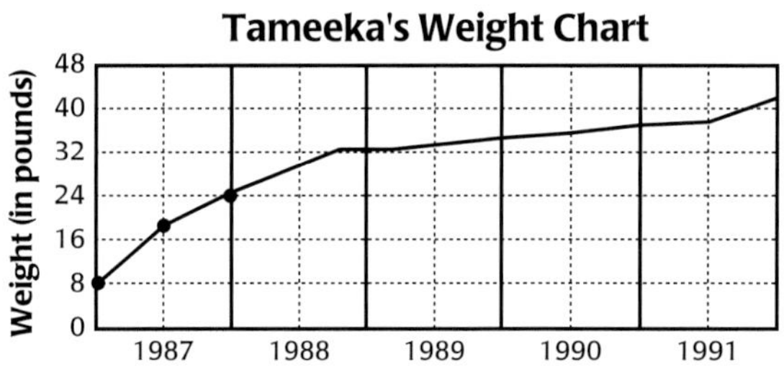

3. a. Answers will vary. Sample response:

Because of her age the monthly fluctuations do not tell much about her weight changes. The doctor will have a better overview during a longer period with this graph.

b. Such detailed information got lost in this graph. You can only compare longer time periods.

c. In 1987 she gained the most weight, about 17 pounds.

Section D. Graphs Depicting Speed

1. Graphs will vary. Sample graph:

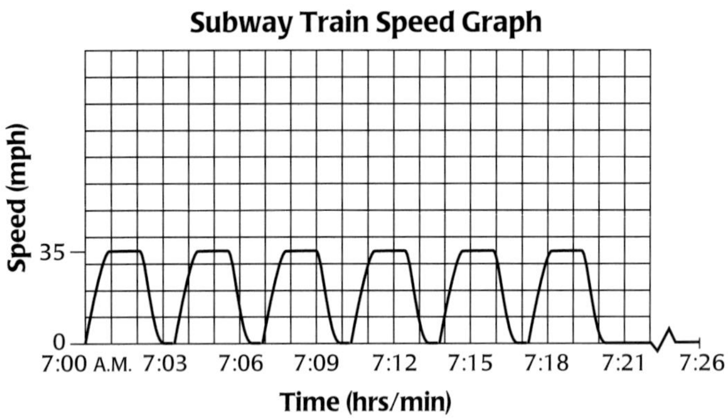

2. Graphs will vary. Sample graph:

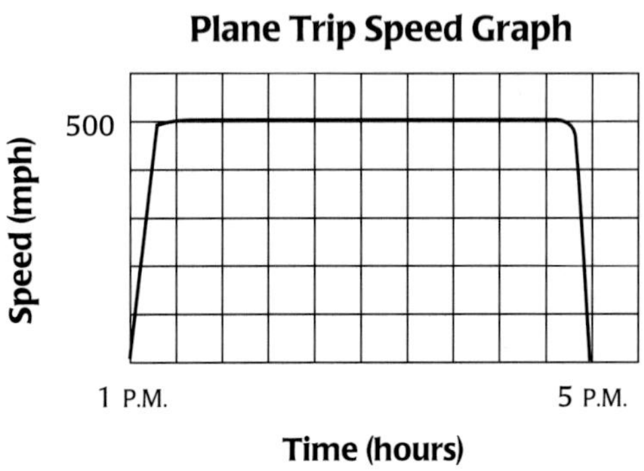

3. Graphs will vary. Sample graph:

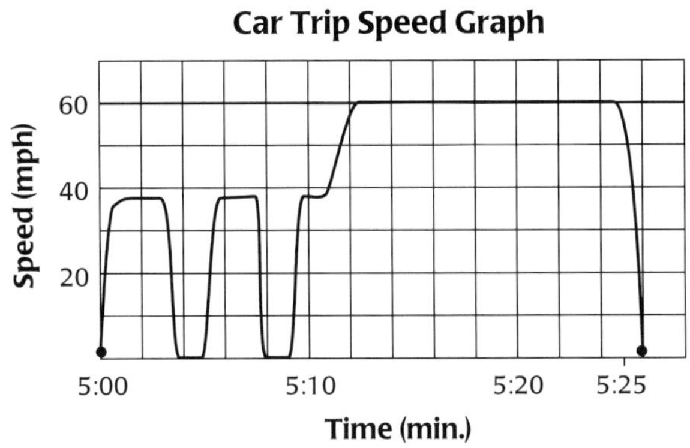

Section E. Graphs of Tides

1. The best fishing occurs around 6 A.M. and 6 P.M.

   The worst times are around midnight and noon.

2. around 6 A.M. and 6 P.M.

3. The tide is receding from 6 A.M. until noon and from 6 P.M. until midnight.

4. Low tide will also be $\frac{1}{2}$ hour earlier than shown on the graph: at 11:30 A.M. and 11:30 P.M.

CREDITS

Cover

Design by Ralph Paquet/Encyclopædia Britannica Educational Corporation.

Collage by Koorosh Jamalpur/KJ Graphics.

Title Page

Phil Geib/Encyclopædia Britannica Educational Corporation.

Illustrations

12, 14, 16 Phil Geib/Encyclopædia Britannica Educational Corporation; **18** Brent Cardillo/Encyclopædia Britannica Educational Corporation; **24** Phil Geib/Encyclopædia Britannica Educational Corporation; **28** Phil Geib and Paul Tucker/Encyclopædia Britannica Educational Corporation; **36** Paul Tucker/Encyclopædia Britannica Educational Corporation; **42, 56, 68, 70, 72** Phil Geib/Encyclopædia Britannica Educational Corporation; **80, 82** Tom Zielinski/Encyclopædia Britannica Educational Corporation; **84, 86** Phil Geib/Encyclopædia Britannica Educational Corporation.

Photographs

6 © Focus on Sports Inc.; **8, 10** © ALLSPORT PHOTOGRAPHY (USA) INC.; **20** © John P. Kelly/The Image Bank; **78** © Jim Strawser/Grant Heilman Photography, Inc.; **80** © Cameramann International.